Oceanic Wilderness

Warty Frogfish, *Antennarius maculatus*, Anilao, Philippines.

Oceanic Wilderness

Mysteries of the Deep

Roger Steene

NEW HOLLAND

This edition first published in 2006 by
New Holland Publishers (UK) Ltd
Garfield House
86–88 Edgware Road
London W2 2EA
United Kingdom

www.newhollandpublishers.com

ISBN 10: 1 84330 912 2
ISBN 13: 978 1 8433 0912 3

Published and produced in Australia by
Crawford House Publishing Australia
14 Dryandra Drive
Belair SA 5052
Australia

www.crawfordhouse.com.au

Digital reproduction by
the Buearu
135 Gilles Street, Adelaide SA 5000
www.digitalprintaustralia.com

Printed by Thomson Press (India) Ltd.

Captions previous pages
pages 2-3, Aerial, Palau.
pages 4-5, Amphlette Is., Papua New Guinea.
pages 6-7, Exposed corals, Great Barrier Reef, Australia.
pages 8-9, Madang, Papua New Guinea.
pages 10-11, Reef scene, Fiji.

For Jerry, Scott, Tono, and Rob

Thank you

This book is dedicated to those who assisted most in its production. Dr Jerry Allen of Perth, Australia; Scott Michael of Nebraska, USA; Takamasa Tonozuka of Bali, Indonesia, and Rob Vanderloos of Alotau, Papua New Guinea.

My old buddy Dr Jerry Allen still manages to join me on field trips despite a waning interest in underwater photography that has been replaced by a passion for mountaineering. Our times together are always enjoyable and productive and he remains ready and prepared to assist when called on. I wish to thank him for his valuable input into the book.

Scott Michael was my constant diving companion over the five years of fieldwork at localities as diverse as the Florida Keys, West Pacific, Caribbean, temperate Australia, Japan and Indonesia. His knowledge, enthusiasm, good humour, and willingness to share made the field trips a delight. I also thank him for his special help on the computer and his untiring efforts to assist me with difficult shore entries.

Takamasa Tonozuka (Tono) of Dive and Dive's, Bali, figured prominently in my endeavours to capture rare animals on film from both deep water and his Secret Bay hideaway. We spent numerous hours inspecting new dive sites both night and day, and shared many wonderful meals in the course of our endeavours. I thank him also for being a wonderful host.

Rob Vanderloos of Milne Bay Marine Charters, Papua New Guinea, was instrumental in my obtaining some of the more recent photos. The facilities on his boat *Chertan* and easy diving resulted in me doing nine trips with him in just over a year. He has impressive local knowledge and is one of the best spotters in the business. It is not unusual for him to do five or six dives a day with his video camera and he invariably spends more time underwater than his guests.

The area I consider one of the best for macro photography is Lembeh Strait in northern Sulawesi, Indonesia. Mark Ecenbarger of the Kungkungan Bay Resort has enabled me to dive this area continuously since my first visit in 1992. The excellent facilities make this one place I never tire of revisiting. Trained local dive guides locate photo subjects for guests, an innovation I wholeheartedly support as eyesight diminishes.

Rudie H. Kuiter always made himself available for fish discussions and shared camera subjects and dives. He has a prodigious knowledge of electronics and has fitted custom systems to my housings, which unfortunately will never be available to the public.

Toshikawa and Junko Kozawa of Anthis Corp., Japan, organized field trips to Osezaki and Oshima Island. I wish to thank them for their efforts and for showing us their top dive sites. I use and enthusiastically endorse their superb *Nexus* housings.

Peter, Suzi, and Chris Parks of Image Quest 3-D, UK, were of great help in my efforts to obtain plankton and microscopic photos at the Lizard Island Research Station on the Great Barrier Reef. A special thank you for allowing me the use of their wonderful optical bench.

In my home town, Nick Tonks (Reef Centre Fuji Image Plaza, Cairns, Australia) and Fenton Walsh provided invaluable assistance. Both donated considerable time, effort and resources to the project.

Neville Coleman contributed support particularly with discussions and identifications of the molluscs.

Annemarie, Helmold, and Danja Köhler kindly invited me on uncrowded live-aboard field trips in Papua New Guinea, as did Kandy Kendall. My gratitude to Ray Izumi, whose untrained eye can detect even the most insignificant grammatical error.

Special thanks to my colleagues who provided additional photographic material. Dr Mary Jane Adams, Clay Bryce, Neville Coleman, Craig Dewit, Rudie H. Kuiter, Scott Michael, Denise Neilsen-Tackett, Rob Vanderloos and Fenton Walsh.

I acknowledge with grateful thanks the assistance also provided by Dr Peter Arnold, Brian Bailey, Dr Ruth Barnich, Rod Barrel and Cathy Holloway (Nai'a Charters, Fiji), Dr and Mrs. H. Batuna (Murex Dive Resort, Manado, Indonesia), Tyme Baumtiken, Fred Bavendam, Dr Penny Berents, Dr Sandy Bruce, Robert Burn, Dr Stephen Cairns, Jimmy Chonga, Dr Pat Colin and Lori Bell-Colin (Coral Reef Research Foundation, Palau), C. Cumberque, Dr Peter Davie, Steve Drogin, Bisquee duBois, Connie Ella, Dr Katharina Fabricius, Julian Finn, Dr Mark Fleming, Zübe Fondelage, Larry Gann, Dr Karen Gowlett-Holmes, John Greenamyer, Bob Halstead, Dr Douglas Hoese, Bob and Ronda Hollis (Oceanic Products, San Leandro, USA), Dr Pat Hutchings, Dr Barry Hutchins, John Jackson, Larry Jackson, Peter Jennings, Lea Jensen, Joe Kennedy, Dik Knight (Loloata Island Resort, Port Moresby, Papua New Guinea), Dr Jeff Leis, Dr Jim Lowry, Peo and Maleta Luke, Andrea Marshall, Woody Mayhew, Dr Patricia Mather, Terry McCassie, Dr Laurence McCook, Dr David and Pamela McKowen, Janine and Shatheed Michael, Dave Miller, Dr Gary Morgan, Dr Mark Norman, Dr Leslie Newman, Drs. John and Hannelore Paxton, Lotta Pfuffage, Gomez O. Podd, Pierce Pott, Shirley, Sandy and Kim Quested, Pang Quong, Dr John E. Randall, Bruce Robison, Denise Rosen, David Salmanowitz, Emiko Shibuya (Dive Paradise Tulamben, Bali, Indonesia), David Staples, Dr Walter Starck, Lyle Snr., Bev, Lyle Jnr., and Cadel Squire, Larry and Denise Tackett, Tawali Resort, Milne Bay, Papua New Guinea, Dr Harry ten Hove, Bill Tewes (Dive St. Vincent, West Indies), Miki Tonozuka, Mark Turdette, Dr Lyle Vail and Dr Anne Hoggett (Lizard Island Research Station, Australia), Cherie Vanderloos, Boy Venus (Club Ocellaris, Anilao, Philippines), Dr Charlie Veron, Annie Wanfer-Sachs, Stan Waterman, Dr Tim Werner (Conservation International, Washington, D.C.,USA), and Dr Fred Wells.

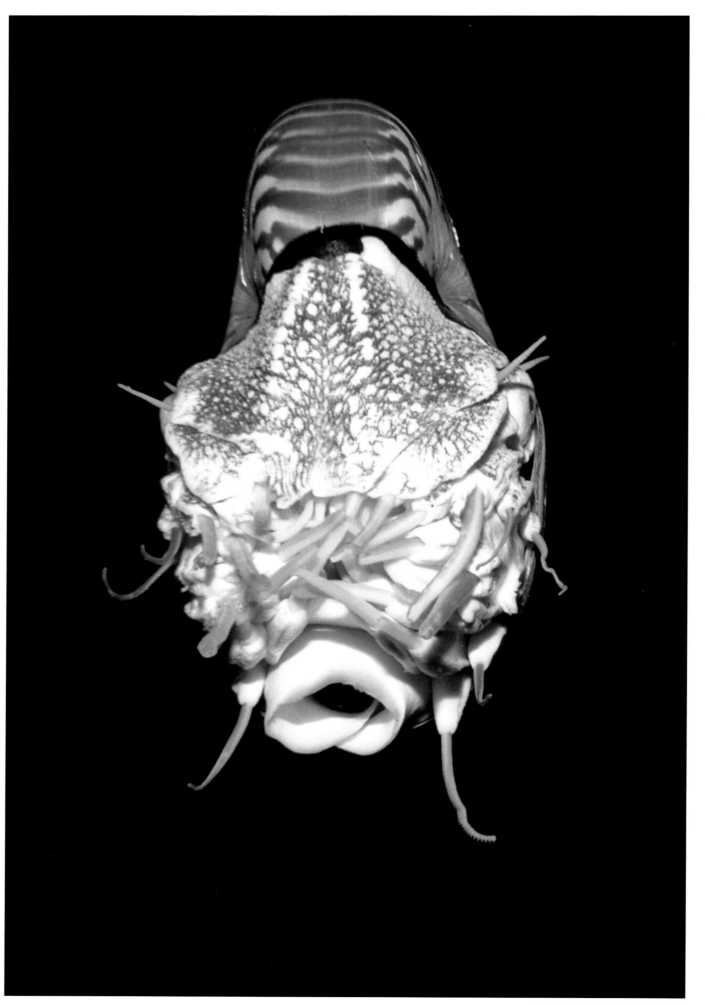

Emperor Nautilus, *Nautilus pompilius*, Russell Is., Solomon Is.

Introduction

Biodiversity. That is the first word that comes to mind when I think of a coral reef. Coral reefs are home to an incredible array of marine organisms. The reef structure itself is comprised of hundreds of different types of plants and animals. There are thousands of algae and sessile or encrusting invertebrates that form a living veneer over much of the reef's framework. There are literally thousands of motile invertebrates and fishes that live on and move over the reef. Many of these shelter among nooks and crannies. Then there are the microscopic as well as large pelagic animals that float or swim immediately above. Thousands of different microscopic plants and invertebrates are at the mercy of currents that affect the reef, while hundreds of different fish species cruise the water column in search of a meal.

It is this diversity that originally attracted me to coral reefs. Anyone who has donned mask and snorkel and stuck their head under the surface of tropical seas is taken aback by sheer numbers, as well as the beauty of the coral reef. There are few ecosystems where you can see hundreds of species during a single 60-minute foray (the average amount of time devoted to a single dive using scuba). The only ecosystem that rivals the coral reef is the tropical rainforest, where many of the animals are less active and more cryptic, at least during the day.

Some of my favourite habitats are those adjacent to coral reefs. This includes the sea grass meadows, mangrove swamps, rubble tracts, and sand or mud slopes. I am particularly fond of the latter two. Here one can find a special community of marine animals that are specifically adapted to these more structurally impoverished environments. Diving on sand or mud slopes is often referred to as muck diving. It has increased in popularity as more experienced divers and underwater photographers

continually hunt for the rare and bizarre. Soft substrates provide a great habitat in which to find sea anemones, sea slugs, cephalopods, crustaceans, snake eels, frogfishes, sea horses, ghost pipefishes, scorpionfishes, stargazers, dragonets, flatfishes and puffers. I have found that black sand substrates, present around volcanic islands, are a real bonanza for underwater photographers. Not so much that there are different species on this substrate, but those animals that are there tend to be more boldly coloured. A good example of this phenomenon are the shrimp gobies. My best shrimp goby photos have come from the black sand beaches off Bali.

The animals that live on or near the coral reefs exhibit an amazing array of adaptations. Perhaps one of the most extreme examples is seen in the flatfishes, many of which reside on the sandy bottoms adjacent to coral reefs. Their young look similar to the larvae of many other bony fishes. This includes the placement of the eyes – in the larvae there is one eye on each side of the head. But some time after the young fish hatches, and before its assumes its benthic lifestyle, one of the eyes slowly begins to migrate. It moves over the top of the skull so that both eyes are on one side of the head. Either the right or left eye will migrate, depending on the flatfish group or species in question. For those species that lie on their left side (e.g., the right-eyed flounders) the eye moves onto the right side of the head. In the left-eyed flounders, both eyes are on the left side of the head. In some flatfish species, either eye may migrate. Not only does one of the eyes shift position, the jaws often become twisted towards the upper side of the body and one pectoral fin (the one on the exposed side) gets larger than the other.

One of the principle adaptations any species must possess in

Coral bommie, Great Barrier Reef, Australia.

order to survive is a way to avoid being eaten. Some marine creatures have evolved deadly toxins to thwart the attacks of predators and to capture their prey. The toxins possessed by some marine organisms are much more potent than any venomous snake. Even some of the most benign looking creatures can do severe damage to a human. A scientist I know who collects marine organisms for cancer research once told me about an unidentified blue sea anemone he found off the east coast of Australia. Upon touching the animal, he was almost knocked unconscious by the sting, even through the gloves he was wearing!

Among the best known members of Australia's venomous marine fauna are the blue-ringed octopuses. According to cephalopod expert, Dr Mark Norman, there are at least five species living around the coast (several of these are undescribed). These beautiful little cephalopods use their venom, which is produced by their saliva glands, to kill the crabs that they feed on. In some cases, their prey are larger than they are and can do considerable damage with their stout claws, so it is necessary to have a very potent weapon to disable them quickly. On very rare occasions, blue-ringed octopuses will bite individuals when they are provoked. One such case was related to me by an Australian invertebrate expert, Dr Karen Gowlett-Holmes. She knew a

person that was bitten by a Southern Blue-ringed Octopus (*Hapalochlaena maculosa*) when he accidentally squashed the animal under his wrist. He had to be kept awake for several days because every time he would drift off to sleep he would stop breathing. She said that he was still suffering affects from the neurotoxic venom six months after the incident. Even though these animals possess such a deadly weapon, they are still vulnerable. A dive buddy of mine witnessed a large Australian cuttlefish (*Sepia apama*) eat a Southern Blue-ringed Octopus.

There are other invertebrates that are distasteful, poisonous or that sting potential predators. Certain sea slugs are well known for incorporating the stinging cells of their hydroid prey into the appendages on their backs. There are also nudibranchs that utilize biological weapons. Some produce sulphuric acid in glands in the mantle that they release when agitated. Some may also absorb toxins found in the sponges they feed on which make them unpalatable to most predators.

An attack by a predator does not always end in death to the prey species. I once saw two-thirds of a jack swim past that had no doubt survived a predator's attack. The fish, which was part of a large school, had its tail and part of its body bitten off. Somehow, the wound healed and the handicapped fish had survived. The

Red Sea divers, Egypt.

poison fanged blennies (genus *Meiacanthus*) often survive being ingested by larger fishes. When consumed, these blennies quickly respond by biting the inside of their attacker's mouth with their poison fangs. The predator usually spits out the blenny, which swims off while the larger fish yawns and breaths rapidly as result of irritation caused by the bite. Predators quickly learn to avoid *Meiacanthus* blennies and several cardinalfishes that mimic them.

Some species of invertebrates and fishes are able to change their colour and body texture in order to better avoid detection by predators. One group of reef fishes whose ability to adapt to their surroundings has given ichthyologists fits, are the ghost pipefishes (*Solenostomus* spp.). There are a number of supposed species, like the Rough-nosed Ghost Pipefish and Halimeda Ghost Pipefish that may actually be ecological variants of one of the more common forms. A dive colleague who lives on the island of Mabul conducted experiments to determine how much a ghost pipefish can change in appearance. He moved a Robust Ghost Pipefish from a sandy area to one with lots of algae and monitored the fish for months. It began to grow dermal filaments all over its body, transforming from smooth to hairy! It is likely that even the highly sought after Red Setter Ghost Pipefish, which is often found in association with red filamentous algae, is

just a variant of one of the more common species. It has simply grown red dermal filaments to blend in better with the algae.

Some of the most fascinating marine animal adaptations involve prey capture and reproduction. Take the apparently harmless and cumbersome looking Murex. Its beautiful shell, with its comb-like extensions, is highly sought after by collectors. Apparently, this relatively benign looking creature can use these spines to entrap its mollusc prey. Murex have also been seen to congregate in groups to lay eggs in a single mass. Hermit crabs are quick to detect the scent of the newly released soft eggs and gather in large numbers to feed. Ingeniously, the shells huddle tightly around the egg mass making a thorny protective cage around the eggs that the hermits cannot penetrate. In a few hours the egg cases harden and the scent disappears. The Murex abandon the eggs, which at that stage are of no further attraction to hermits or other predators. The hardened cases remain undisturbed in the open until hatching.

There are some specialized fishes that angle for their prey. The best known of these are the frogfishes (or anglerfishes). They have a modified first dorsal spine that is used to lure prey into striking range. The esca (the lure) takes on many different forms, depending on the species in question. There are some that

Unknown sea slug, Milne Bay, Papua New Guinea.

have lures that look like tiny mysid shrimp or amphipods. The Striated Frogfish (*Antennarius striatus*) has a lure that looks like a swimming polychaete worm. While the esca of this species serves primarily as a visual attractant, it also has secretory cells in the lure that are thought to emit an enticing scent. There have been Hispid Frogfish (*Antennarius hispidus*) collected in deep-water whose esca glowed in the dark. These individuals had bioluminescent bacteria living in this structure. The walking batfishes, the Decoy Scorpionfish (*Iracundus signifer*), and the stargazers also have specially adapted body parts that they use to attract their prey. There is also a flounder, *Asterorhombus fijiensis*, that has a modified dorsal spine complete with a rod and a lure.

Reef creatures also display some amazing reproductive adaptations. Many marine invertebrates and fishes are hermaphrodites, that is, they can change sex or possess both sex organs simultaneously. Nudibranchs are well-known simultaneous hermaphrodites. They have tube-like genital apertures on the right side of their bodies. When they mate, they adopt a head-to-tail orientation and the two animals exchange sperm packets through the everted reproductive structures.

There are also many hermaphroditic fishes. Most reef

fishes change sex from female to male. These are known as protogynous hermaphrodites, and include certain morays, anthias, wrasses and parrotfishes. There are a smaller number of fishes that are protandric hermaphrodites involving male to female sex change. The most popular species in this group are the ribbon eels and the anemonefishes. A relatively small number of fishes are simultaneous hermaphrodites. Morays and groupers exhibit this reproductive mode. Finally, there are both-direction sex changers. That is, they can change from female to male, and then back again if the social situation should change (e.g., there are too many males). Certain coral-dwelling gobies are known to change sex in both directions.

The hamlets (*Hypoplectrus* spp.), a genus of small Atlantic groupers, are simultaneous hermaphrodites that display particularly interesting reproductive behaviour known as egg trading. When a pair of hamlets meet to spawn at dusk, they will alternate sexual roles. One fish is the male first, while the other is the female. They will then change roles and spawn again. Studies have shown that it is preferential to begin the nightly spawning ritual as a male. In some cases, individuals will refuse to play by the rules and always attempt to be males (release sperm). However, these individuals are avoided in the future by potential mates.

Wavy Bubble, *Micromelo undulatus*, Milne Bay, Papua New Guinea.

A relatively small number of reef fishes engage in some parental care of their eggs or larvae. The most basic kind of care consists of the male, female or both parents defending a benthic nesting site. Male dottybacks are known to chase potential egg predators away from the egg ball that the female deposits in a crevice. In egg-tending species, not only is the tending individual responsible for egg defence, they will often fan the eggs with their fins or remove eggs from the "nest" that are showing signs of fungal infection.

One of the best known groups of egg tenders are the jawfishes. These burrow-dwelling fishes orally incubate their eggs, holding them in their mouths until they hatch. The Threespot Frogfish (*Lophiocharon trisignatus*) is a lesser known egg-tender. It carries a cluster of eggs on the side of its body. The relatively large eggs, which can number as many as 650 in a batch, are attached to the male fish's body by thread-like structures. Not only does this reproductive mode ensure that the eggs are more likely to survive, it also enhances the parent's chances of catching a meal. Opportunistic fishes that are attracted to the eggs on this highly camouflaged species turn from potential egg predator to frogfish prey when they get too close to the brooding parent.

The Spiny Chromis (*Acanthochromis polyacanthus*) is one of a handful of reef fishes that are known to take care of their young as well as the eggs. After hatching, the larvae form a cloud that hangs around the parents. They have even been reported to feed on the slime of the adult fish by nibbling on the body surface.

The most incredible example of parental care in marine fishes is exhibited by the Banggai Cardinalfish (*Pterapogon kauderni*) from Indonesia. All of the cardinalfishes are known to orally incubate their eggs. After the female lays an egg mass, the male ingests them. He holds them in his mouth for about one week until they hatch, at which time the larvae enter the plankton. But in the case of the Banggai Cardinalfish, not only are the eggs orally incubated, the newly hatched young are taken into the male fish's mouth where they develop for another 13 to 18 days. By the time they leave the male, they are tiny replicas of the adult fish. Although exhibited by a number of freshwater cichlid fishes, this is the only species of marine fish that is known to mouth brood its young.

My friend Dr Jerry Allen and I originally discovered this fascinating reproductive mode by travelling to Banggai Island to see this fish. At that time, it was only known from this location. In the last few years, it was also discovered in Lembeh Strait. I

Yellow Seaperch, *Lutjanus monostigma*, Sharm-el-Sheikh, Egypt.

had spent much time looking for this fish in northern Sulawesi, but I had never seen one until 1999. So why did the fish suddenly show-up here? Apparently it was accidentally (or possibly intentionally) introduced to the Strait by a local fish collector. Since 1999, it has moved from a couple, to many of the popular dive sites in the area. Fortunately, it does not appear to be competing for microhabitat with any indigenous cardinalfishes. It is found both among sea urchin spines and with sea anemones. In fact, it will actually swim among the tentacles and make contact with those sea anemone that have less potent stinging cells, like the Leathery Sea Anemone (*Heteractis crispa*).

I am often asked by young divers I meet during my travels, "wasn't it better back when you first started diving?" I always answer this question with a resounding "No!" As recently as 20 years ago, it was difficult to reach many of the places I am now able to get to with relative ease. In my early diving days, it was a challenge to dive the Great Barrier Reef, which is at my own back door. We would have to hire a boat and make the long run out to the inner or middle reef for the day. It was lots of effort for relatively little bottom time. Now there are dozens of live-aboard dive boats that work this same area. You can go out on a comfortable vessel and spend days or even weeks exploring and photographing on the outer reef and Coral Sea. The same applies

to places like Malaysia, Indonesia, Papua New Guinea and the Caribbean. It is now easier than ever before to dive in relatively remote areas.

Despite the ease with which photographers can get to exotic dive locations, it is still possible these days to find unusual subjects and record undocumented behaviour. Take the case of the Mimic Octopus, featured in this book. In the late 1980's my companion Rudie Kuiter came ashore after a dive on the Indonesian Island of Flores and said he had just seen an octopus on open sand that was mimicking a flounder. We quickly went back out to witness the most remarkable marine animal behaviour it has been my privilege to observe. Here was a beautifully marked octopus, periodically emerging from holes in the sand to travel over a large arc in search of food. As it did so, its form would change from that of an octopus to a stingray. It would then stand motionless and turn into a sand anemone. Each time it returned to a hole, it would swim up off the bottom replicating a flounder. When safely back in a hole, the head would appear from the sand as a snake eel. To this day, I have not seen another Mimic Octopus perform as spectacularly as this.

On returning to Australia, I related my experience to Dr Anne Hoggett at the Lizard Island Research Station on the Great

Yellowfin Goatfish, *Mulloidichthys vanicolensis*, Sipadan I., Malaysia.

Barrier Reef. That jogged her memory of an incident she had seen also involving an octopus and mimicry. A large tightly packed school of parrotfish was swimming from one high coral head to another, across an expansive open sand gutter. One of the fishes at the bottom of the group caught her eye, for some reason it just looked different. As the school finally crossed the open area and reached the safety of the far coral head, the fish that had caught Dr Hoggett's eye slowly separated from the others, turned into an octopus and dropped down onto the coral. Obviously it had used its incredible skills to cross safely from one reef to another. This was not the same species we had found at Flores but yet another octopus with the ability to change shape, form and motion at will. I have since recorded two additional octopus species also engaging in intelligent mimicry.

Even though the presence of the Mimic Octopus has been known for years, little is understood about the animal. It is difficult to find photos or footage of its incredible antics. To achieve this, you might have to spend long hours of observation. It does the mimicry when it feels like it, not when the photographer wants it. Also individuals differ. Some do the mimicry often, others don't. It does not always happen like my first unforgettable sighting in Flores. Young and juveniles, like children, have to hone these skills as their lives progress and are mostly unproductive for photographic purposes.

A good example involved a film crew that went to Indonesia to record the Mimic Octopus in action. After spending all of a week observing a solitary individual, they compiled a documentary stating that multiple mimicry was a myth and the octopus only mimicked a flounder. Subsequently I worked with a B.B.C. film unit, and despite being well prepared and equipped, it was only on the last four days of an intensive six weeks expedition including diving around the clock, that interesting mimicry was filmed. It took me years of patience to compile the photos of this species of octopus appearing in the book.

Unfortunately still photographs cannot document the relevant motions the animal uses to accompany its different mimicry. To complicate matters there exists a smaller, more beautiful but very similar octopus (an unnamed species) that is mostly mistaken for the true mimic. This animal can also be found in the same areas but has an almost non- existent repertoire of mimicry. Perhaps one day a budding young researcher will find the time to study the life cycle of the true Mimic Octopus and record just what this amazing creature is capable of.

Another factor that makes diving better in the year 2004, is the access to marine life information. When I first started diving, there was very little popular literature available on marine

Leafy Filefish, *Chaetodermis pencilligera*, Lembeh Strait, Indonesia.

organisms. But now the crop of high quality books on marine life is incredible! This is especially true in the case of fish books, although invertebrates guides are also becoming more readily available. No doubt within the next 10 years there will even be guide books available on some of the invertebrate groups whose members are more difficult to identify, like sponges, soft corals, and tunicates. To give some indication of how many books are now available on marine animals, consider Neville Coleman in Australia. He just published his 56th book relating to marine animals and the marine environment. Amazingly, four of his most recent publications feature a staggering 6000-plus colour photographs.

As a result of the proliferation of these types of books, as well as the increased number of educational television documentaries, the modern diver is far more informed and knowledgeable than we were back when I first started diving. In the early days, we were terrified of sharks, all sharks. Most were considered to be man-eaters, that would attack a diver whenever the opportunity presented itself. Now, thanks to the efforts of many field biologists and sport divers, we know that sharks rarely bite and when they do, it may be motivated by threat not hunger. Things have changed so much for the better, that the fearsome Great White Shark (*Carcharodon carcharias*) is now protected in

California, Australian and South African waters.

Many sport divers are making important contributions to our knowledge of marine animal behaviour. To give an example, on a recent field trip to Papua New Guinea, a fellow traveller, Ray Izumi, noticed interesting behaviour by the Harlequin Shrimp (*Hymenocera picta*). Izumi is an accomplished videographer who specializes in filming small creatures. He filmed the Harlequin Shrimp using the spikes on its clawed appendages to repeatedly stab the sea stars it was eating. Why they do this is a mystery, but it may be that they are actually puncturing the water vascular system of the sea star by poking holes in it. This would interfere with the animal's ability to move away from the predator.

One of my favourite places to stay is Tulamben, Bali. The reason I like this place is because at the end of the day, divers sit around in the restaurant and share their day's observations. An interesting story related to me by an enthusiastic young lady involved moray behaviour. She was snorkeling when she saw a pair of morays in apparent combat. The eels were actually biting one another and both bore gaping wounds as a result of the melee. Suddenly, the larger of the two eels grabbed the smaller moray and began to swallow it. The smaller eel responded by rapidly knotting its tail around the rear of its own body (something

Netted Pufferfish, *Canthigaster solandri*, Lomaiviti Group, Fiji.

that morays often do when they tear chunks of flesh from larger prey items). This prevented the larger eel from swallowing the knotted eel. The larger moray had no choice but to spit the smaller eel out. When it did, both morays swam off, wounded but able to fight again another day.

Another Tulamben snorkeler shared an interesting observation, again involving morays. He saw a circle of fish around a low coral head that included opportunistic predators, like groupers, goatfishes and wrasses, but also some parrotfishes. Eventually he noticed what was attracting the roving band of fishes. There were two morays fighting at the base of the coral head. The battle ended when one of the morays bit the other one in half. What was interesting, is that the curious onlookers did not swoop in to feed on the damaged carcass, but instead they dispersed once the fighting was over. Were the fish just interested in being spectators to a fight, like boys in a schoolyard? No, these predators were probably taking advantage of the small fishes, crustaceans and other potential food flushed out of hiding by the moray battle. The parrotfishes would not have been attending the party to capture fleeing prey, but they would eat sponges that grow under pieces of disturbed coral rubble.

Although we have definitely come a long way in our knowledge of the marine environment and its inhabitants, there are still amazing discoveries being made every year. A new species of coelacanth was revealed in the Manado area of northern Sulawesi in 1997. American biologist, Dr Mark Erdman and his wife Arnaz, were visiting a local fish market when they found a bizarre fish they immediately recognized as the fabled coelacanth. However, they were not aware that the fish had never been reported from Indonesia so they simply took a photo of the specimen and did not purchase it from the fisherman. The photos were eventually seen by coelacanth expert Dr Eugene Balon and the search for another specimen began. In 1998, Erdman was successful in acquiring a second coelacanth. After in-depth study, it has been confirmed that the Indonesian coelacanth represents a different species that has been named *Latimeria menadoensis*. The locals call the fish Rajah Laut, which means "King of the Sea" and say the oily flesh is good for curing problems with the gastrointestinal tract.

Not only are there marine animals that have yet to be discovered, there are regions of the world that have yet to be explored. Even in heavily utilized regions like the Caribbean, there are still special places that see relatively little diver traffic. In 1999 I visited the island of St. Vincent. What an amazing place! The quantity of unusual marine animals seemed more reminiscent of the western Pacific than the other places I had visited in the Caribbean before.

Round Crab, *Demania splendida*, Lembeh Strait, Indonesia.

The western Pacific is still the best place to find a new dive site or discover a new fish or invertebrate species. Perhaps the world's best diving has yet to be discovered in a massive area we call the "coral triangle," an area that includes the Philippines, Indonesia and Papua New Guinea. The waters of Indonesia boast the richest marine fauna in the world. At the seldom visited Raja Ampat Islands off western New Guinea, ichthyologist Dr Jerry Allen recorded 286 fish species at one site, the most ever recorded anywhere in the world on a single dive! Likewise, on the same expedition, coral guru Dr Charlie Veron surpassed all previous coral counts. Not only are the Raja Ampats and the surrounding area an underwater paradise, they are also stunningly beautiful above the waterline. I was amazed to see incredible beehive islands there, more impressive in size and quantity than the fabled rock islands of Palau.

As one moves farther from Indonesia, marine animal diversity gradually declines. An estimated 2,700 species of fishes have been reported from the inshore waters of Indonesia. In comparison, the Society Islands have around 630 species, while 566 species have been reported from the well studied inshore waters of the Hawaiian Islands. There are far fewer new coral reef fishes being discovered than when I first started travelling with Dr Allen and Dr John Randall, the world's foremost authorities

on tropical reef fishes. Nevertheless, there are still new species being described all the time. In fact, in the year 2001, 132 new marine fishes were named.

Many of the species being found today are from relatively poorly known deep reef habitats. This habitat, from a depth of 65 to 160 metres (200 to 500 feet) is called the Twilight Zone. To explore these areas, divers are employing sophisticated diving techniques using rebreathers to go deeper and spend less time decompressing. This involves far more risk than shallow water scuba diving. Eric Reichardt, a fish-collector who assisted me during a trip to the Florida Keys, died in November 2001 as a result of a deep water accident. He was diving with a rebreather in 80 metres (250 feet) when he either had an equipment malfunction or was attacked by a shark (his body was found with shark inflicted injuries, but it has yet to be determined if they were suffered before or after his death). Even so, some have chosen to take these risks to increase our knowledge of this poorly known habitat. One of the leading pioneers in Twilight Zone exploration is Richard Pyle of the Bernice Bishop Museum, Hawaii, who has discovered many new deep-dwelling fish species. Dr Patrick Colin has also made amazing discoveries using a one-man submersible that he used to explore deep reefs (to 320 m) off Palau. His efforts have yielded a number of very interesting species, including a rare

Fire Urchin Shrimp, *Allopontonia iaini*, Milne Bay, Papua New Guinea.

butterflyfish and a fantastic new angelfish that has a colour pattern reminiscent of the raccoon butterflyfish (*Chaetodon lunula*).

Although often the bane of environmentalists, marine aquarists have also added to our understanding of the marine environment and its inhabitants. This is especially true in the case of stony corals. Many reef aquarists are growing scleractinians with the same success that people grow house plants. In my travels, I have met aquarists who have literally cropped thousands of stony coral fragments from their tanks to send to fellow reef keepers. Aquarists have made significant observations concerning asexual and sexual reproduction in corals. They have also added to our insights into coral metabolism, growth rates, trace element utilization, competition and aggression. Aquarists have an advantage over field biologists in that they can more easily gather information on individual animals for months or even years.

Fish-keeping marine aquarists have also made contributions to our knowledge of fish behaviour. Scott Michael, author and marine aquarist, documented an amazing feeding behaviour in the Tasselled Wobbegong (*Eucrossorhinus dasypogon*). A juvenile kept in a large home aquarium was observed to use its tail as a lure to attract potential prey. The tail, which looks like a small fish complete with an eye and a caudal fin, is held up and over the head and wagged from side-to-side when food is present!

In recent years, marine aquarists have had much more success in spawning and rearing the larvae of their piscine charges. Almost all the anemonefishes, many of the dottybacks, comets, reef basslets, cardinalfishes, gobies and some of the other smaller reef fishes have been successfully bred and raised in captivity. Not only has the commercial production of some of these species helped to take pressure off wild stocks, it has provided biologists with insight into the reproductive behaviour and larval development of a number of reef fish families. No doubt as time goes by, aquarists will unlock the secrets into the husbandry of some of the larger and less durable marine species, like the butterflyfishes. In fact, one enterprising group has succeeded in rearing the larvae of coral-feeding butterflyfishes by feeding them normal aquarium foods.

Unfortunately, aquarium fish collectors have committed some serious environmental atrocities. The first time I encountered the Psychedelic Mandarinfish (*Synchiropus picturatus*) was at Secret Bay, Bali. There was a large population living among stands of Hammer Coral (*Euphyllia ancora*) in a shallow portion of the bay. Much to my delight, and that of my friend who owns the dive operation there, they were less secretive and easier to work with than their better known cousin *Synchiropus splendidus*. But on my last trip to the bay, I was horrified to find that all the hammer coral had been pulverized and all the mandarins were gone. We

Creole Wrasse, *Clepticus parrae*, Bonaire, Caribbean.

concluded that fish collectors had found out about the presence of these fish, which are highly prized in the aquarium trade, and had come in and destroyed the coral to get to the hiding mandarins.

There has been a dramatic increase in the popularity of diving since I took up the sport in 1964. According to PADI, the largest certification agency in the world, they have certified a total of 10,151,841 divers since 1967. Since PADI reports that they certify one out of every two divers certified worldwide, it is probably safe to assume that more than 20 million divers have been certified in the last 30 years. In the year 2000, about 854,000 individuals were certified worldwide by PADI, while only 3, 226 were certified in 1967. In 1980, PADI certified about 107,000 divers. Although many more than in the late 1960's, there were still in excess of eight times as many divers certified by this agency in 2000 than in 1980! That is a lot of divers.

Doesn't the incredible influx of sport divers put incredible pressure on coral ecosystems? We often think that a large number of divers equates to a dead or at least unhealthy reef. While there is no doubt that divers can cause some aesthetic damage, I have found that areas that experience heavy diver traffic can still be wonderful places to visit. In September 2001,

I had the good fortune to dive Osezaki, Japan. This is the most popular dive site in the country, in part because it is only a few hours by car from Tokyo. On weekends, it was not uncommon to share the relatively small 400 metre bay at the tip of the Ose peninsula with upwards of 3000 other divers. Despite the heavy diver pressure at Osezaki, the reefs and sand slopes were extremely rich and beautiful. In fact, it was one of my most productive trips, resulting in a number of photos that are featured in this book.

There is no doubt *Homo sapiens* have negatively impacted coral reefs in certain parts of the world. There are also times when humans get the blame when a natural event was actually the main contributor to the destruction. A few years ago I was preparing to go to Sumatra. A German professional photographer told me that I should not waste my time as the reefs had been extensively damaged as a result of dynamite and cyanide fish collecting. According to him, the reefs were all dead. He had been there with other divers and was appalled that people still described it as a dive destination. Even after this grim overview, I decided to go to Pulau Weh.

The reef destruction described by my German colleague was evident right away. Obviously something bad had happened on

Tiger Shark, *Galeocerdo cuvier,* Great Barrier Reef, Australia.

these reefs, as there were extensive areas of dead coral. But it was also obvious that in the ensuing years the corals and other sessile invertebrates were making a strong comeback. I discussed the condition of the reef with the owner of the dive operation, Dodent Mahyiddin, an uneducated local man with a village upbringing. I discovered that long before reef conservation became fashionable, Dodent had taken it upon himself to protect the local reefs. He had publicly protested the use of cyanide and was successful at getting it banned from the area. He was also able to stop some of the dynamite fishing. When I talked to him about the destruction the German photographer had seen, he said that this was not the result of destructive fishing practices. He took me into his dive shop and showed me a book where he had religiously recorded daily sea temperatures for 20 years. With this data he was able to demonstrate that a few years ago the water temperature dropped overnight from 29 to 18 degrees Celsius for eight continuous months! This amazing temperature shift is incredibly rare for an Indonesian locality, where stable conditions usually exist. It was obvious that thermal shock killed the corals, not dynamite or cyanide, and the dates of the coral dieback coincided with the huge temperature drop. In some cases, Mother Nature is much more destructive than we humans are.

Undoubtedly one of the major environmental problems facing coral reefs today is rising sea surface temperatures. Stony corals, soft corals and other anthozoans will expel the unicellular algae that lives in their tissues, known as zooxanthellae, if temperatures rise or fall outside of a certain temperature range. This will also happen if they are stressed by other environmental conditions (e.g., sedimentation, increase in inorganic nutrients, disease, excess shade, increase in UV radiation). When these animals expel the zooxanthellae or when there is a reduction in photosynthetic pigments in these plant cells, the animal is said to bleach. It is called bleaching because the coral's white skeleton becomes visible through the now transparent tissues.

The expelling of the algae can cause major problems for the coral. Although they feed by capturing small planktonic organisms in their polyps, the algae also provide an important source of food. The byproducts of the algae photosynthesis supply nutrients for the corals. Therefore, when the algae die, the corals may starve to death. Fortunately, if sea temperatures return to more normal levels within about five to ten weeks, the corals will often regain their symbiotic algae and survive. However, if temperatures remain outside the tolerance range for longer periods of time, the corals will eventually die.

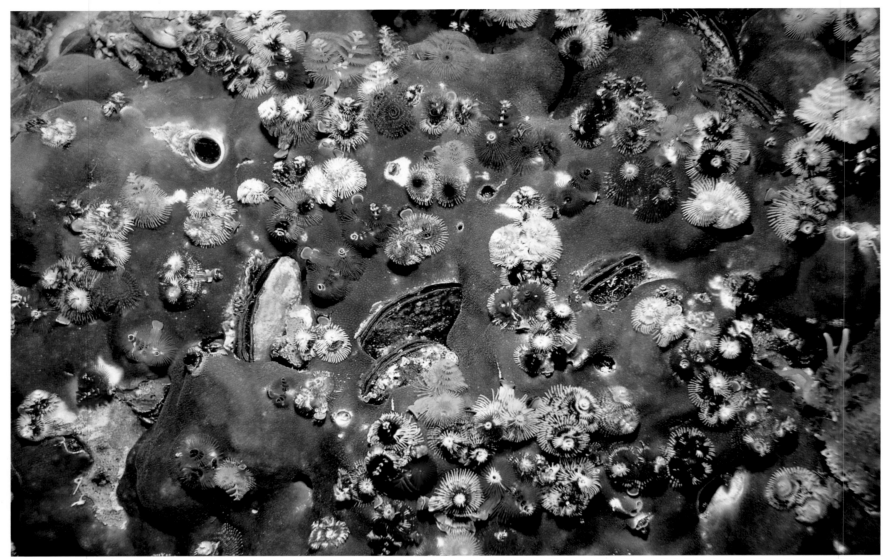

Christmas Tree Worms, *Spirobranchus giganteus*, Semporna, Malaysia.

Unfortunately, the bleaching phenomenon has occurred with greater regularity over larger geographical regions in the past 20 years. It has had a huge impact on diving in the Indo-Pacific in the last five years, with catastrophic bleaching events having occurred in the Maldives, Micronesia, the Great Barrier Reef and numerous other popular dive sites in this region.

While coral reefs are taking a hit as a result of recent warming in sea surface temperatures, their resilience has often been underestimated. I remember during a visit to a popular dive resort in the Caribbean, we were told by the dive-masters that by simply touching a coral head you could kill the entire colony. At this location, they did not even allow divers to wear gloves for fear of the damage that would result. Stony corals are much more durable than that. In the early 1960's when the Australian government was contemplating oil drilling on the Great Barrier Reef, clandestine experiments were conducted to see what the impact of an oil spill would be on stony corals. A floating barrier was constructed around a coral head to allow the monitoring of a controlled spill within the enclosure. When the tide went out, the coral was covered with the viscous, sticky fluid. Over the ensuing weeks and tides, the coral gradually sloughed off its mucus along with the oil, and subsequent monitoring indicated the coral survived. A horrifying scenario, but also a lesson, if a

stony coral might withstand this kind of trauma, it can certainly survive being touched by a human finger.

A question that I am frequently asked is which up and coming photographer's work do I find most impressive? Without singling out an individual, I am convinced that the Japanese underwater photographers are ahead of the rest of the world. All you have to do is pick up a Japanese dive magazine and compare it to similar publications from the USA, Europe or Australia. The photographic subject material is light years ahead of the rest! While we in the western world were concentrating on wide-angle photography, the Japanese perfected their macro-photography equipment and techniques. You can see this trend if you compare the camera gear used by the average photographers. American divers in particular still use large, cumbersome arms and strobes when doing macro-photography. In contrast, the Japanese have compact systems with tiny strobe arms and small, lightweight housings. This enables them to get into tight spaces without bashing the coral and scaring off their subjects.

The Japanese are also very much into marine animal behaviour and do a remarkable job documenting it. It is not uncommon to see a photo spread in a Japanese dive magazine showing the complete life history of a specific species of goby. They are not

Coral tips, *Acropora secale*, Komodo I., Indonesia.

only educated in photographic methodology, they also know their subject matter very well and often specialize in a particular group of fishes or invertebrates. I know a number of Japanese photographers that only shoot gobies. Another individual spends his time hunting out and photographing only damselfishes. These are not professional ichthyologists, but merely amateur photographers that have a passion for these particular families.

One thing my American dive buddy and I noticed during our stay at Osezaki was that the Japanese divers, even those new to the sport, would pore over the immense fish and invertebrate guide library at the resort after every dive. This was not a small percentage of the visiting divers, but almost all of them would find the species they had seen and enter them one by one into their log books! For someone involved in the production of a number of marine life field guides, this was very exciting to see.

I am happy to say that we in the western world are going in this direction. We are learning more about our subjects and spending more time seeking out and photographing the smaller members of the reef community. I think this trend will continue.

The camera gear of today also lends itself to taking better photos. Even the novice can take perfectly exposed, in-focus

photos of a shy invertebrate or nimble fish. You no longer have to be a seasoned veteran to snap that award winning shot. In fact, some of the best photos I have ever seen were taken by relatively inexperienced amateurs that happened to be in the right place at the right time with the right gear.

One thing that I love about the new camera systems is their small size. I have always tried to create the most compact system I can. With this type of gear, it is easier to get into tight spaces to hunt more elusive subjects. Of course, things are going to get even better with the new digital technology. It will not be long until we are using palm-sized camera housings with internal flashes to take amazing photos. I am also looking forward to the day when I don't have to carry hundreds of rolls of film along with me and worry about what the x-ray machine may do to it! These cameras are bound to revolutionize underwater photography.

Another exciting trend that I have seen in the past decade is that more individuals are capturing marine animal behaviour on video because of the availability of small cameras and housings. This enables the amateur to record behaviour that may have never been documented before. These videos will no doubt greatly increase the pool of information available to invertebrate and fish ethologists.

Thorny Oyster, *Spondylus varians*, Milne Bay, Papua New Guinea.

A wonderful consequence of my years of working with other photographers and field researchers is the formation of a cooperative global network of friends and colleagues. We readily give each other advice and share photos. Rather than hide the location of an unusual animal that no other person has photographed, we openly share all information. We also exchange images with each another. If anyone is working on a particular project and needs photos, we open our archives so that person can achieve the best possible quality and coverage. This type of cooperative effort reaps enormous rewards, not only for the photographers concerned, but everyone who appreciates the wonders of the oceanic wilderness.

Sea whips, *Junceela fragilis*, Madang, Papua New Guinea.

The Doughboy Scallop is a common resident of reefs and jetty pylons of temperate Australia. Its true colouration is masked by a layer of living sponge, resulting in a variable livery contrasting with the row of eye spots in the opening of the shell.

Doughboy Scallop, *Chlamys asperrimus* (4-7 cm), in 10 m, Edithburgh, South Australia.

Many crustaceans live commensally on corals, which they use as a place to shelter and a source of food. This spider crab gathers its sustenance from food particles that adhere to its host's polyps.

Spider Crab, *Xenocarcinus tuberculatus* (2 cm), in 22 m, Lembeh Strait, Sulawesi, Indonesia.

Crinoids are prehistoric echinoderms that feed by rapidly beating cilia in channels in their arms. Tiny zooplankton gradually pass through these channels and into the crinoid's mouth. The crinoids have cirri on the bottom of the crown to grasp the reef when at rest or while feeding.

Opposite: *Clarkcomanthus* sp. (5 cm), in 15 m, Tulamben, Bali, Indonesia.
Above: Possibly *Oxymetra* sp. (8 cm), in 10 m, Komodo, Indonesia.

These small, geographically disparate reef fishes have evolved remarkably similar colour combinations. The Royal Gramma of the Caribbean exhibits an amazing similarity in both pattern and habits to the three dottybacks from the western Pacific, although not closely related.

Above top: Magenta Dottyback, *Pseudochromis porphyreus* (5 cm), in 15 m, Ishigaki, Japan.
Above bottom: Royal Gramma, *Gramma loreto* (5 cm), in 10 m, Bonaire, Caribbean.
Opposite top: Two-tone Dottyback, *Pseudochromis paccagnellae* (5 cm), in 15 m, Tulamben, Bali, Indonesia.
Opposite bottom: Diadem Dottyback, *Pseudochromis diadema* (5 cm), in 15 m, Busuanga, Philippines.

Stonefishes are normally difficult to see due to their excellent camouflage colouration. This exceptional individual may have recently shed its skin or its colours are possibly adapted to blend with sponge or algal encrusted backgrounds.

Reef Stonefish, *Synanceia verrucosa* (40 cm), in 8 m, Loloata I., Papua New Guinea.

The Ambon scorpionfish was first described from specimens taken in deepwater trawls. It has since been found at shallow depths at several "muck diving" sites in the west Pacific. Like all scorpionfish, its fin spikes pack a nasty, venomous punch.

Ambon Scorpionfish, *Pteroidichthys amboinensis* (7 cm), in 10 m, Secret Bay, Bali, Indonesia.

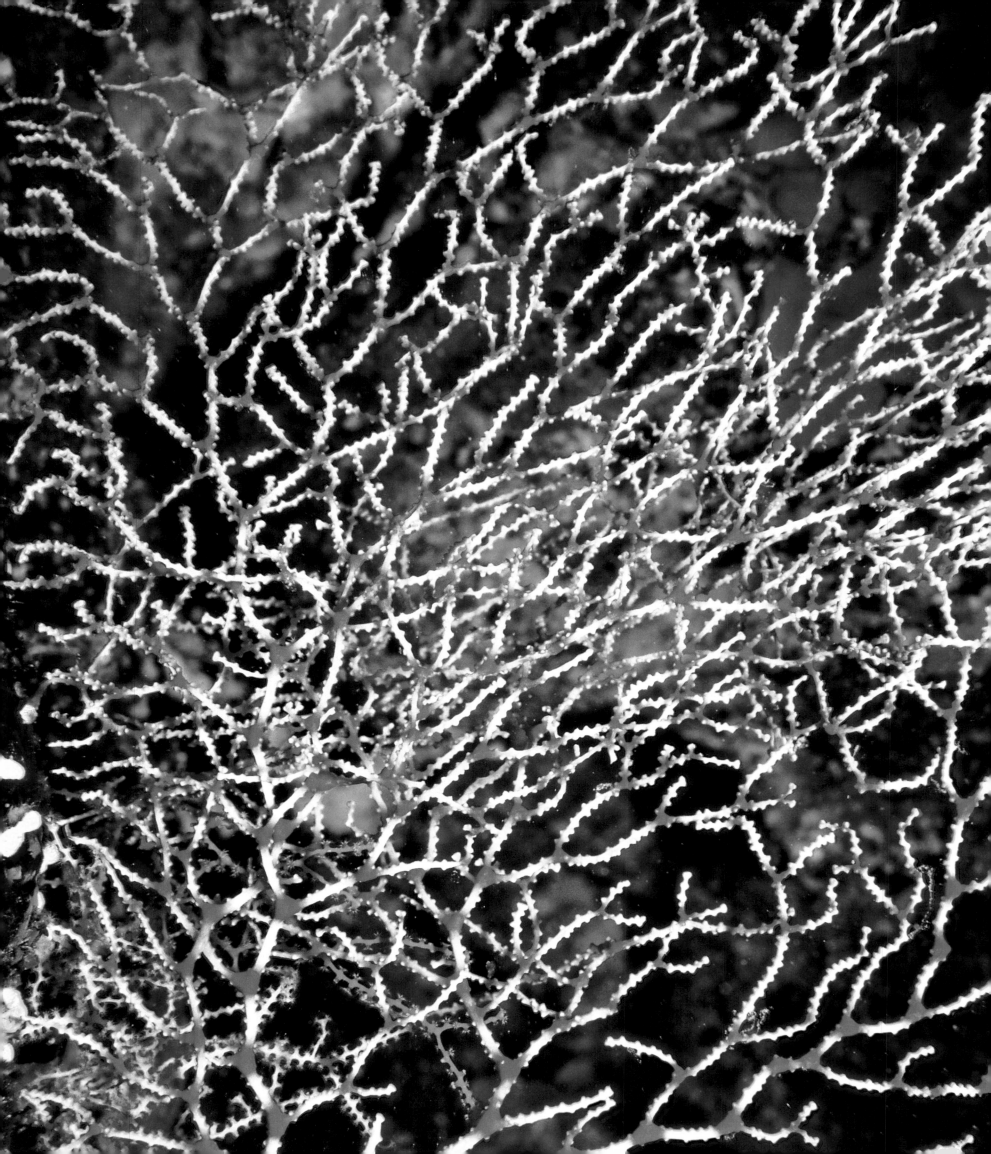

Soft corals form an integral part of the coral reef community, sometimes forming the dominant bottom cover. Their growth forms are tremendously diverse, including slender whips, huge fans, trees, and fleshy colonies.

Opposite: Soft coral, *Acabaria* sp. (20 cm), in 15 m, Sharm-el-Sheikh, Egypt.
Above: Soft Coral, *Scleronephthya* sp. (12 cm), in 10 m, Lembeh Strait, Sulawesi, Indonesia.

Like hard corals, octocorals (soft corals), are comprised of colonies of numerous individuals. Many species, such as those shown here have branching growth forms, an advantage for exposing maximum surface area while feeding on microzooplankton.

Opposite: Soft corals, *Dendronephthya* sp. (20 cm), in 20 m, Russell I., Solomon Is. and Loloata I., Papua New Guinea
Above: Soft coral, *Acanthogorgia* sp. (9 cm), in 18 m, Komodo, Indonesia

Opposite: The supporting structure of some soft corals consists of a maze of sclerites formed of calcium carbonate. They vary greatly in shape and colour according to species and are useful for identification by specialists.

Soft coral: *Stereonephthya* sp. (10 cm), in 20 m, Bintan I., Indonesia.

Above: Pineconefish have patches of skin near the edges of the mouth that act as growing substrate for bioluminescent bacteria. These microorganisms glow green in the dark and are used like headlights by the pineconefish to locate food at night.

Pineconefish, *Monocentris japonica* (3 cm), in 16 m, Osezaki, Japan.

Above and opposite: The symbiotic relationship between this sessile ctenophore and starfish is poorly understood. Apparently the ctenophore, (detail opposite) which divides and spreads by the process of fragmentation, uses the starfish as a mobile feeding platform. It is unknown whether this association is parasitic or simply harmless.

Starfish Ctenophore, *Coeloplana astericola* (2 cm), in 18 m, Tulamben, Bali, Indonesia.

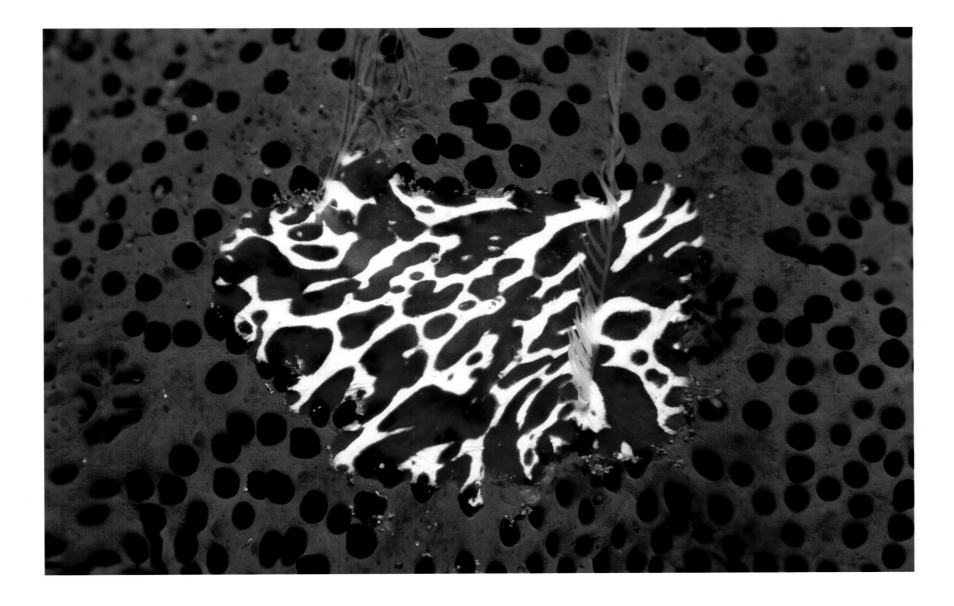

Above: Coleman's Shrimp make their home among the venomous spines of the Fire Urchin. They clear an area of spines and tube feet and this becomes the centre of their entire universe. The urchin can host more than one generation during its life cycle. The smaller shrimp is the male.

Coleman's Shrimp, *Periclimenes colemani* (2 cm). Fire Urchin, *Asthenosoma varium*, in 18 m, Ambon, Indonesia.

Opposite: The acrobatic stance struck by this young Warty Frogfish is a prelude to the emptying of its stomach contents. This is not unique behaviour, members of the scorpionfish genus *Rhinopias* are also known to act likewise.

Warty Frogfish, *Antennarius maculatus* (3cm), in 6 m, Lembeh Strait, Sulawesi, Indonesia.

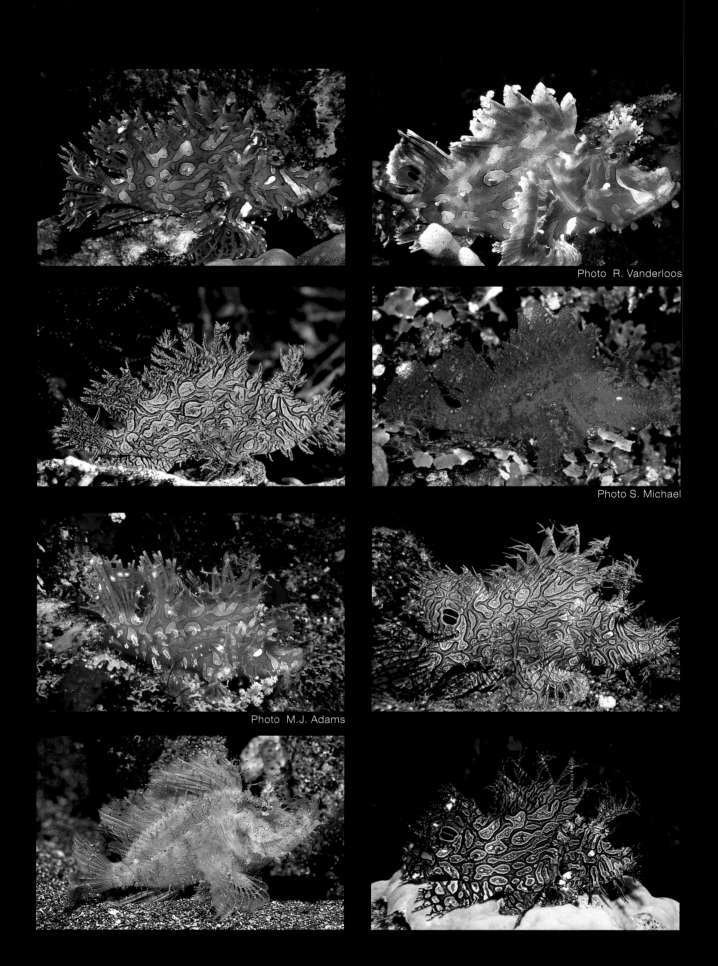

Photo R. Vanderloos

Photo S. Michael

Photo M.J. Adams

Above and opposite: *Rhinopias* scorpionfishes have become a favourite subject of underwater photographers in the Indo-Pacific region. Although venomous and having incredible camouflage, they occasionally fall prey to large reef cuttlefish.

Species include: Merlet's Scorpionfish, *Rhinopias aphanes*; Weedy Scorpionfish, *Rhinopias frondosa*; Eschmeyer's Scorpionfish, *Rhinopias eschmeyeri*; unidentified scorpionfish, *Rhinopias* sp. (5-20 cm), in 8-35 m, Osezaki, Japan; Lembeh Strait and Ambon, Indonesia; Loloata I., Milne Bay, Kavieng and Eastern Fields, Papua New Guinea.

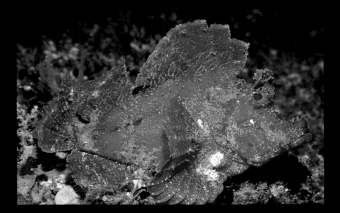

Photo D. Neilsen-Tackett

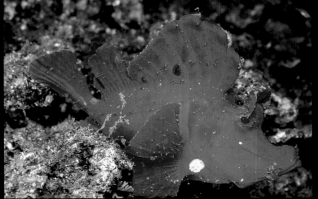

Photo D. Neilsen-Tackett

Photo N. Coleman

Photo C. Dewit

Tunicates filter-feed on fine planktonic particles from the currents. Some attain advantageous feeding positions by generating their own stalks, others may utilize skeletons of dead organisms such as sea fans and sea whips.

Opposite top: Ascidians, *Rhopalaea crassa* (7cm), in 8 m, Lembeh Strait,Sulawesi, Indonesia.
Opposite bottom, *Clavelina detorta* (10 cm), in 12 m, Menjangan I., Bali, Indonesia.
Above, *Perophora namei* (12 cm), in 6 m, Raja Ampat Is., Indonesia.

Opposite: A pair of young Scrawled Filefish adopts a head down posture as they drift just above the sea floor. It appears as though they may be attempting to imitate floating plant material.

Scrawled Filefish, *Alutera scripta* (30 cm), in 8 m, West Palm Beach, Florida, USA.

Above: Lush stands of Sargassum weed often grow in sheltered waters near shallow coral habitats. They provide shelter and feeding grounds for a select fish and invertebrate community.

Sargassum weed, *Sargassum* sp. (1.5 m), in 1 m, Raja Ampat Is., Indonesia.

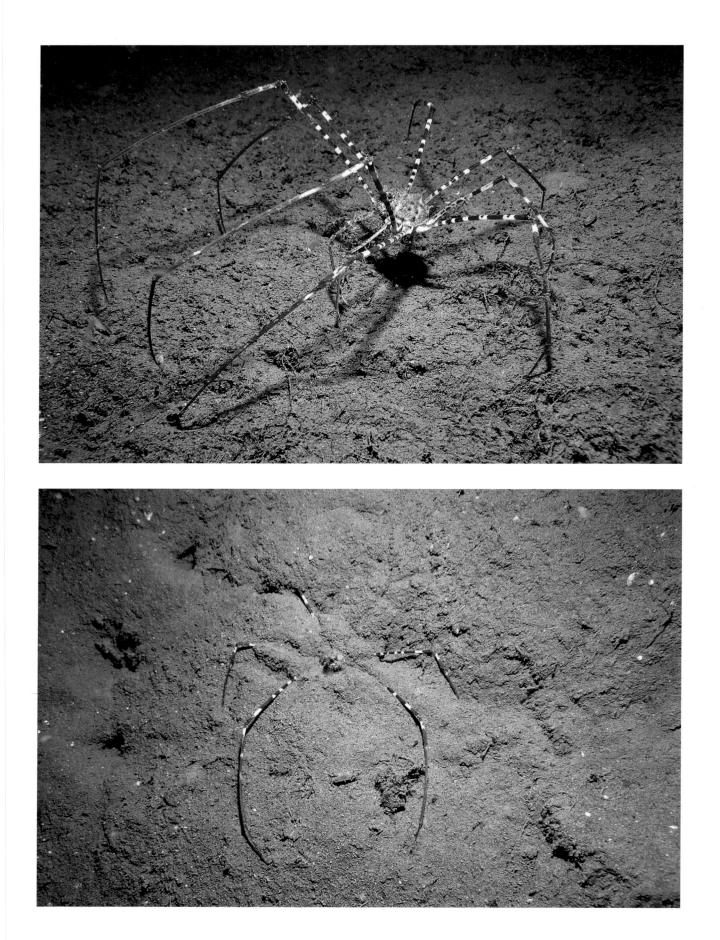

Inhabitants of soft, sandy or muddy substrates are well adapted for life in this seemingly barren environment. Some construct permanent burrows, but many others can quickly bury themselves whenever shelter is required.

Above: Daddy Longlegs Crab, *Latreilla velida* (20 cm),
Opposite: Venus Comb Murex, *Murex pectin* (12 cm), in 10-18 m, Milne Bay, Papua New Guinea.

Previously unknown, this giant scaleworm was discovered by the author and named during the preparation of this book. It lives in a tube secreted by glands spinning the golden threads on its sides with benthic sediments. The feathery appendages (palps) are sensory feeding structures and the flexible scales facilitate water flow for respiration.

Giant scaleworm, Polyodontes vanderloosi, (55 cm), in 7 m, Milne Bay, Papua New Guinea.

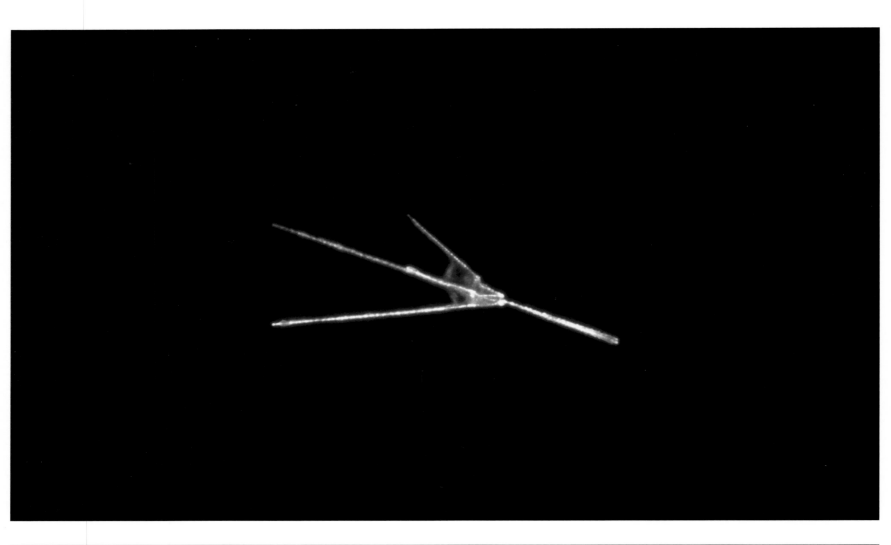

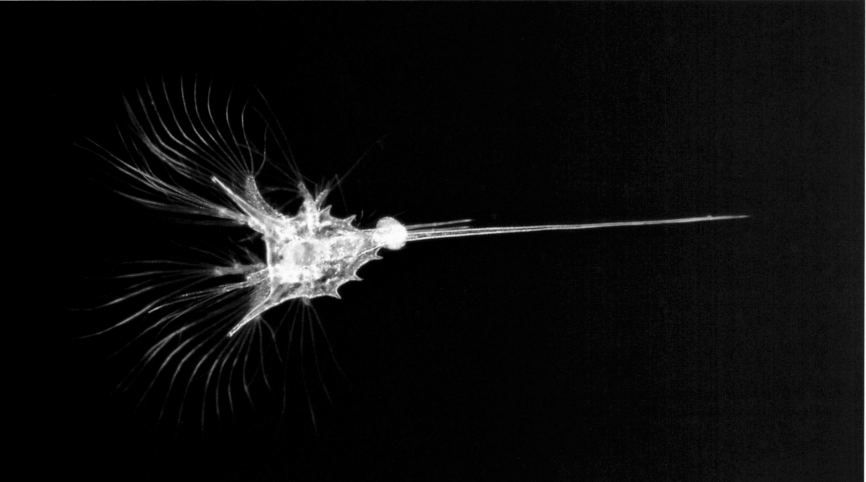

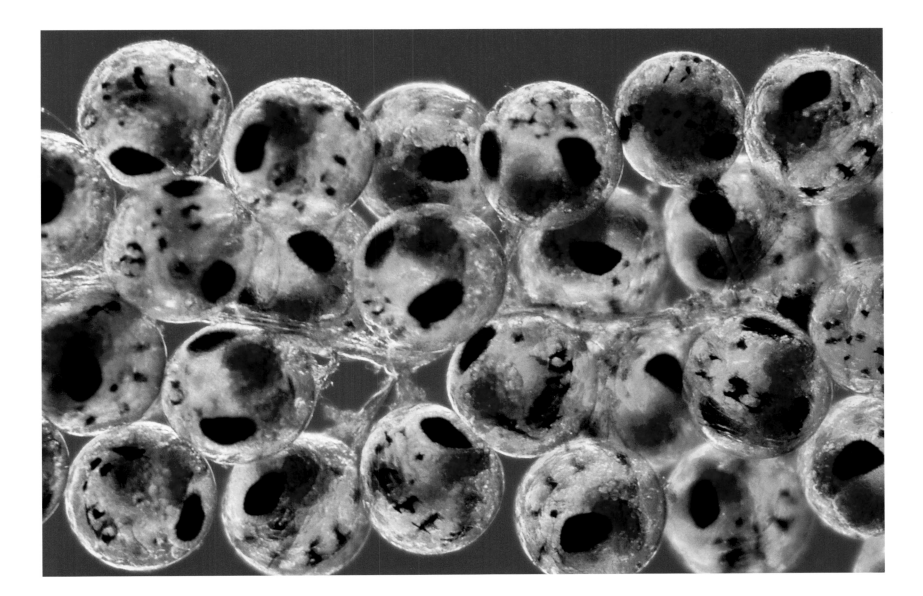

Opposite: The microscopic world of plankton is inhabited by an array of surreal organisms, including the young of many species that bear little resemblance to their adult counterparts. Good examples are this echinopluteus larva of a heart urchin and nauplius larva of a goose barnacle.

Opposite top: Heart urchin larva, unidentified.
Opposite bottom: Goose barnacle larva, unidentified, Great Barrier Reef, Australia.

Above: These crab eggs are approximately three days away from hatching. They emerge as tiny zoea that are pelagic for several weeks. Eventually they settle onto the substrate and metamorphose into young crabs.

Crab eggs, *Cardisoma carniflex,* Great Barrier Reef, Australia.

There is an array of reef organisms that engage in mimicry to dissuade would-be predators. These two cowry species bear a close resemblance to sea slugs that are distasteful to other marine animals.

Above top: Egg Cowry, *Ovula ovum* (4 cm), in 20 m, Lembeh Strait, Sulawesi, Indonesia.
Above bottom: Sea slug, *Phyllidia* sp. (6 cm), in 10 m, Raja Ampat Is., Indonesia.

Top: White Cowry, *Ovula costellata* (5 cm), in 20 m, Lembeh **Strait,** **S**ulawesi, Indonesia.
Bottom: Nudibranch, *Halgerda batangas (*4 cm), in 12 m, Anilao, **Phi**lippines.

Above: The Short-tailed Pipefish is found on sand, mud or rubble bottoms to 25 m, often near sea grass. Widespread in the Indo-West Pacific, it is a stiff bodied species growing to 40cm, the largest encountered by divers in tropical waters.

Short-tailed Pipefish, *Trachyrhampus bicoarctatus* (32cm), in 18 m, Lembeh Strait, Sulawesi, Indonesia.

Opposite: Ghost pipefish present fish taxonomists with a real challenge. At present, no one is really sure how many species there are because of the incredible variability that they display. Their colour and texture can change radically to better blend with their environs.

Ghost pipefish, *Solenostomus* sp. (10 cm), in 8 m, Osezaki, Japan.

Opposite: This biscuit star is a temperate species, ranging along the southern coast of Australia. It inhabits open coastlines and sheltered bays, feeding on ascidians, bryozoans, sponges, and algae.

Southern Biscuit Star, *Tosia australis* (4 cm), in 5 m, Edithburgh, South Australia.

Above: Many nocturnal fishes are equipped with large, sensitive eyes that are effective at extremely low light levels. The Bigeyes have a reflective layer of cells behind the retina known as the tapida lucidium. This structure is responsible for the incredible eye shine seen in these fishes.

Short Bigeye, *Pristigenys alta* (8 cm), in 40 m, Florida Keys, USA.

The Komodo region of Indonesia is renowned for the abundance of this brilliant holothurian. Its feathery arms are tucked in when resting, but fully extend when feeding. Zooplankton is conveyed to the mouth as each individual arm is retracted.

Sea Apple, *Pseudocolchirus violaceus* (15 cm), in 10 m, Komodo, Indonesia.

Found in all marine environments, the crustaceans represent a diverse group of animals second only in number to the molluscs. Mostly active at night, they include a multitude of weird and wonderful members that are relatively easy to observe and photograph.

Top: Red-banded Lobster, *Justitia longimanus* (15 cm), in 30 m, St, Vincent, Caribbean.
Bottom: Paron Shrimp, *Gelastocaris paronae* (1.5 cm), in 5 m, Milne Bay, Papua New Guinea.

Top: Pencil Crab, *Ixa* sp. (4cm), in 8 m, Milne Bay, Papua New Guinea.
Bottom: Round Crab, *Carpilius convexus* (12 cm), in 12 m, Milne Bay, Papua New Guinea.

Shortfin Lionfish, *Dendrochirus brachypterus* (10 cm)

Indonesian Lionfish, *Pterois kodipungi* (20 cm)

Japanese Lionfish, *Pterois lunulata* (18 cm)

Pygmy Lionfish, *Brachypterois serrulata* (10 cm)

Volitans Lionfish, *Pterois volitans* (20 cm)

Twinspot Lionfish, *Dendrochirus biocellatus* (10 cm)

Hawaiian Lionfish, *Dendrochirus barberi* (7 cm), (photo S. Michael)

Longfin Lionfish, *Pterois* sp. (15 cm)

Lovely yet lethal, the lionfishes are armed with venomous spines to ward off predators. The enlarged pectoral fins are used to block the escape route of their prey as these fishes slowly approach to within striking range. There are approximately 20 species worldwide.

In 8-25 m, Lembeh Strait, Secret Bay, Manado, Weh Island, Indonesia; Christmas Island, Indian Ocean; Osezaki, Japan; Milne Bay, Papua New Guinea; Hawaii; Red Sea.

Spotfin Lionfish, *Pterois antennata* (15 cm)

Bluefin Lionfish, *Parapterois heterura* (12 cm)

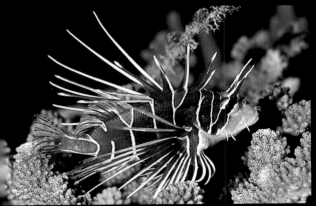

Russell's Lionfish, *Pterois russelli* (20 cm)

Deepwater Lionfish, *Pterois mombasae* (12 cm)

Clearfin Lionfish, *Pterois radiata* (10 cm)

Zebra Lionfish, *Dendrochirus zebra* (10 cm)

Indian Ocean Lionfish, *Pterois muricatum* (18 cm)

Bellus Lionfish, *Dendrochirus bellus* (8 cm), (photo R.H. Kuiter

The alternating dark and light banding pattern exhibited by a diversity of marine organisms may function to disrupt their outlines and shapes. Some species also mimic the banded colour pattern of certain venomous sea snakes.

Above: Brownbanded Bamboo Shark, *Chiloscyllium punctatum* (22 cm),
Opposite: Ornate Octopus, *Octopus* sp. (45 cm), in 5-18 m, Lembeh Strait, Sulawesi, Indonesia.

Opposite: This larvacean (pelagic sea squirt) drifts in ocean currents within an elaborate mucous "house" with filtration grids and escape tunnels. Visible here with the aid of a chemical dye, it can be abandoned if necessary, and another rebuilt in an hour.

Larvacean, *Oikopleura* sp. (1 cm), in 4 m, Great Barrier Reef, Australia.

Above: This odd structure is the sand encrusted tube built by the aptly named Radar Tubeworm. Found in areas of strong water movement, the radial filaments act as a trap to catch current borne deposits on which this terebellid feeds.

Radar Tubeworm, *Lanice* sp. (4 cm), in 15m, Milne Bay, Papua New Guinea.

Above and opposite: The stinging tentacles and leathery skin of a tube anemone offer a protective home for cleaner shrimps in open areas. As well as servicing fishes, these shrimps also gain sustenance from foraging and gleaning planktonic deposits from the anemone.

Tube anemone, *Cerianthus* sp.,Cleaner Shrimp, *Periclimenes* sp., (4 cm), in 20 m, Milne Bay, Papua New Guinea.

The brilliant and sometimes psychedelic mantles of these tridacnid clams act as algae farms. Endosymbiotic algae photosynthesize, providing nutrients and oxygen to their clam host. The colour pigments in the clam's mantle act as a sunscreen, blocking harmful UV radiation.

Burrowing Clam, *Tridacna crocea* (8-15 cm), in 2-10 m, Indonesia and Papua New Guinea.

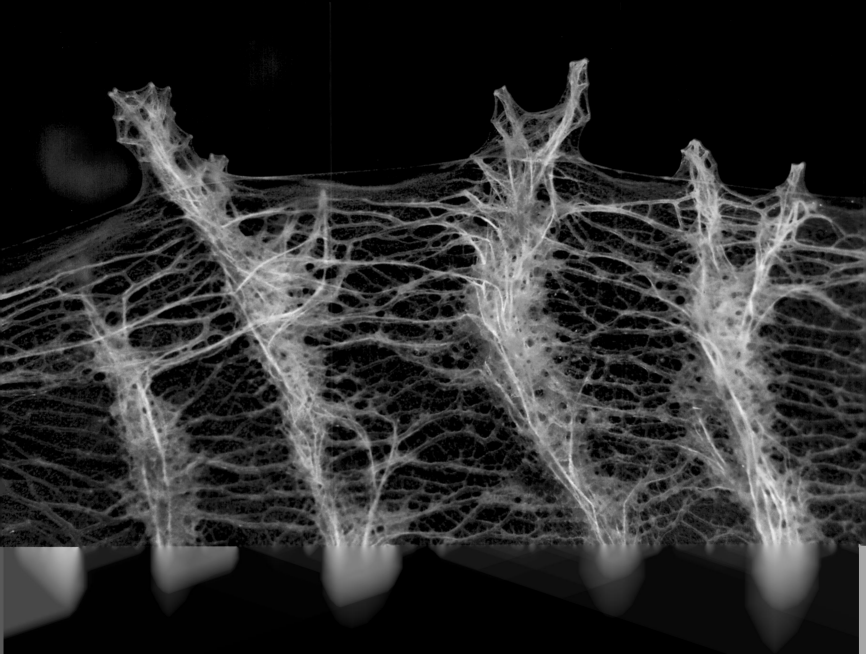

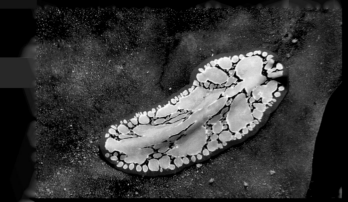

Pseudoceros cf. *scriptus* (3 cm)

Pseudoceros imperatus (4 cm)

Pseudoceros lindae (4 cm)

Pseudoceros laingensis (5 cm)

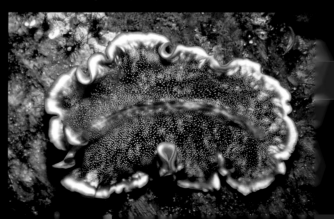

Pseudoceros dimidatus (6 cm)

Pseudoceros rubroanus (5 cm)

Thysanozoon sp. (5cm)

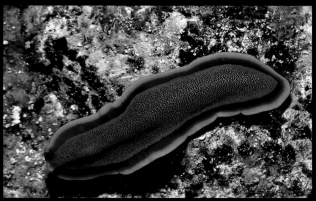

Pseudoceros ferrugineus (4 cm)

Pseudoceros imitatus (5 cm)

Phrikoceros baibaiye (4 cm)

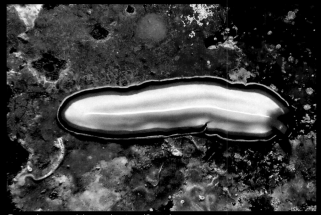

Pseudoceros bimarginatus (3 cm)

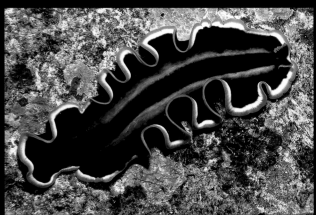

Pseudobiceros sp. (5 cm)

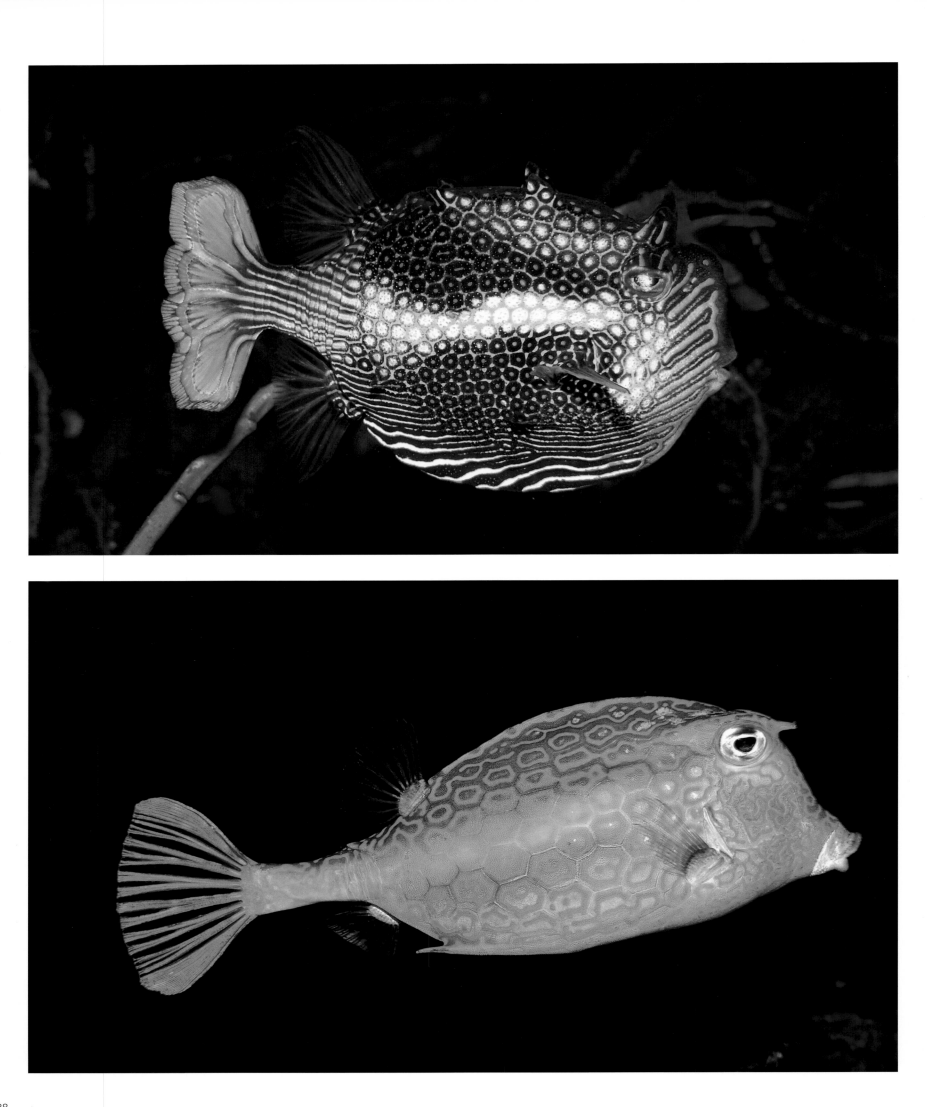

Opposite: The bodies of the horn-bearing cowfishes are covered with bony armour. Not only are they armour plated, but they also exude a toxic slime. These two features protect them from predators.

Top: Ornate Cowfish, *Aracana ornata* (12 cm), in 10 m, Edithburgh, South Australia.
Bottom: Honeycomb Cowfish, *Lactophrys polygonia* (15 cm), in 15 m, Bonaire, Caribbean.

Above: The catsharks comprise one of the most species rich shark families. Many are secretive, hiding in reef crevices, under rocks and benthic debris. This gulf catshark was found under a waterlogged piece of timber, curled up and snoozing like a house cat.

Gulf Catshark, *Asymbolus vincenti* (60 cm), in 6 m, Edithburgh, South Australia.

Opposite: Some nudibranchs are carnivorous and feed on other opisthobranchs. Although capable of swallowing small prey whole, this Red-bump nudi will rasp away with its radula until the flap in its mouth is consumed. The victim, being too large to swallow, will probably then escape.

Red-bump Nudibranch, *Gymnodoris rubropapulosa* (7 cm), Purple-edge Ceratosoma, *Ceratosoma tenue*, in 20 m, Milne Bay, Papua New Guinea.

Above: Approximately one fish in 1000 makes it through the planktonic stage to settle on the reef but even then survival is a risk. It must run the ever-present gauntlet of benthic predators among which the Lizardfishes are most active.

Reef Lizardfish, *Synodus variegatus* (12 cm), in 15 m, Milne Bay, Papua New Guinea.

Above and opposite: Camouflage is the primary defence of many crab species. This male-female pair, although differing in appearance arrives at a similar deception. The male has attached corkscrew algae to spikes on its rostrum while the female relies on remarkable shape and form to deter predators.

Spider crabs, *Huenia heraldica* (2-3 cm), in 5 m, Milne Bay, Papua New Guinea.

Opposite: The colour patterns of many juvenile angelfishes are different than those of the adults. By sporting disparate attire, juveniles can settle out of the plankton and establish a small territory within the domain of the adult without getting picked on by their more mature relative.

Opposite top: Blue-ringed Angelfish, *Pomacanthus annularis* (5 cm), in 12 m, Bintan Is., Indonesia.
Opposite bottom: Emperor Angelfish, *Pomacanthus imperator* (6 cm), in 10 m, Milne Bay, Papua New Guinea.

Above: The fins of some fishes are adorned with streamers. Mostly they disappear as the fish matures, but one exception is the Longtail Dartfish. Adults have been recorded with tails equal in length to the body of the fish.

Longtail Dartfish, *Ptereleotris hanae* (15 cm), in 15 m, Osezaki, Japan.

Above: These small, mobile tube-dwelling amphipods are related to other microcrustaceans (isopods and mysids). Their distinctive tubes are constructed from sea grass, algae, or detritus. Young hatch as mini-adults without a larval stage.

Amphipod, possibly *Cerapus* sp. (1 cm), in 10 m, Milne Bay, Papua New Guinea.

Opposite: The aberrant pattern evident on these hard corals is known as a neoplasm. It is essentially a genetic form of "cancer" capable of reproducing a separate "species" that is distinct from the parent colony. Further research is required before this phenomenon, common on all reefs, is clearly understood.

Top and bottom: Brain Coral, *Diploria strigosa* (field of view 10 cm), in 6-10m Florida Keys, USA and Bonaire, Caribbean.

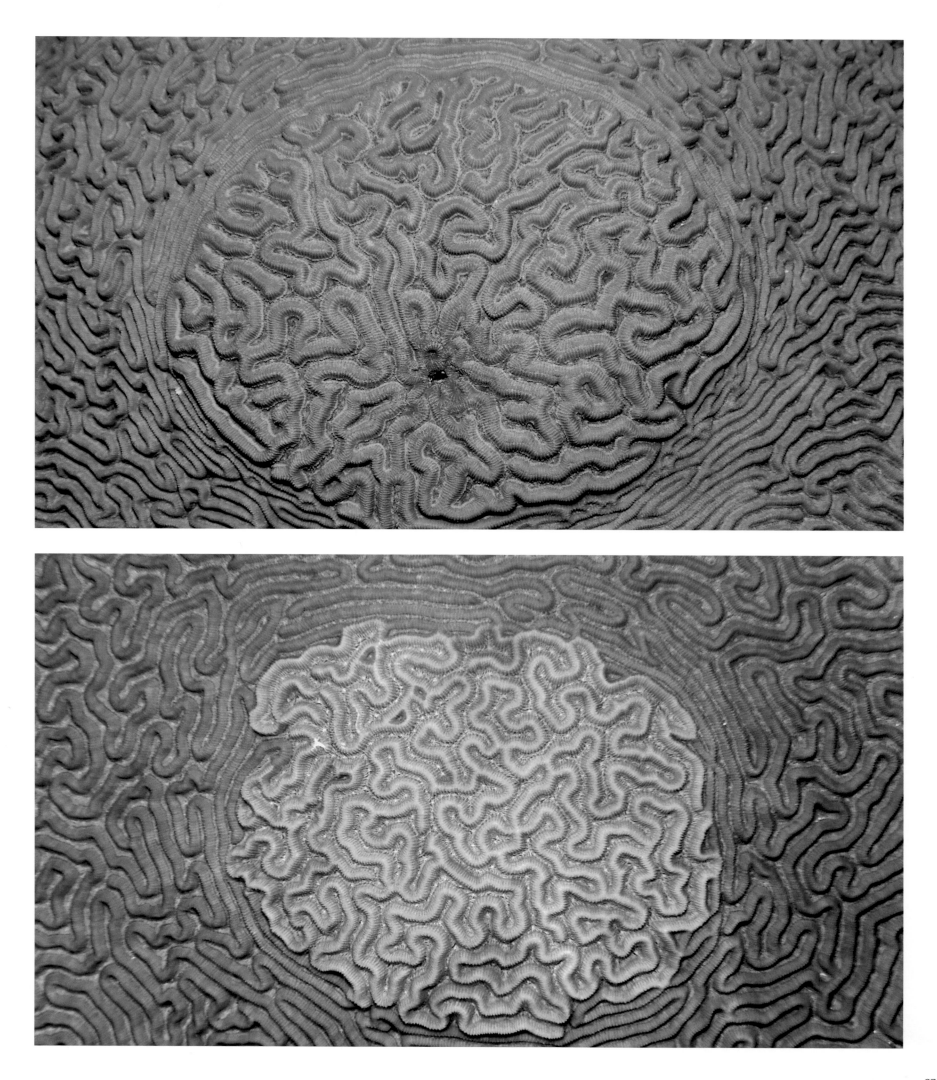

The common and scientific name of this ascidian is derived from its brain-like appearance,
although other species may exhibit a range of shapes. Colonies are comprised of individual zooids
interconnected by a firm jelly-like matrix. Colours are highly variable.

Brain Ascidian, *Sycozoa cerebriformis* (6 cm), in 5 m, Spencer Gulf, South Australia.

The thorny oyster has simple eyes that consist of a cornea, a lens and a retina. With these sensory organs they can detect sudden changes in light levels, snapping shut in a instant.

Thorny Oyster, *Spondylus varians* (18-20 cm), in 6-10 m, Milne Bay, Papua New Guinea.

Nudibranchs have both male and female sex functions. They copulate by aligning and connecting everted genital organs through which the reciprocal transfer of sperm is made. Each individual produces its own eggs, which after hatching, undergo a larval planktonic stage before final settling.

Above: Red-gill Nudibranch, *Nembrotha megalocera* (5cm),
Opposite top: Tail-gate Nudibranch, *Risbecia tryoni* (5cm),
Opposite bottom: Crested Nudibranch, *Nembrotha cristata* (6cm),
in 3-12m, Lembeh Strait, Sulawesi, Indonesia.

Above and opposite: Vermetid snails construct worm-like tubes that are often cloaked in colourful encrusting sponge. They are immobile, and like oysters and barnacles, permanently attached. They feed on microscopic plants and animals in surrounding currents.

Vermetid snail, Vermetidae (10 cm), in 10 m, Raja Ampat Is. and Lembeh Strait, Sulawesi, Indonesia.

Some crabs exhibit remarkable camouflage. This species actually snips polyps from its soft coral host and attaches them to its carapace, changing them as the fragments of coral eventually wilt.

Dendronephthya Crab, *Hoplophrys oatesii* (2 cm), soft coral, *Dendronephthya* sp., in 12 m, Milne Bay, Papua New Guinea.

Some of the commensal shrimp are masters of disguise, perfectly matching the shape and colour of their hosts. The Keyhole Shrimp hangs upside down on sea whips, corals and other vertical structures.

Keyhole Shrimp, *Angasia* sp. (4-5 cm), in 15-25m, Milne Bay, Papua New Guinea and Lembeh Strait, Sulawesi, Indonesia.

Milne Bay Demoiselle, *Chrysiptera cymatilis* (4 cm)

South Seas Devil, *Chrysiptera taupou* (5 cm)

Goldtail Demoiselle, *Chrysiptera parasema* (4 cm)

Goldbelly Damsel, *Pomacentrus auriventris* (5 cm)

Yellowfin Damsel, *Chrysiptera flavipinnis* (4 cm)

Goldtail Demoiselle, *Chrysiptera parasema* (4 cm)

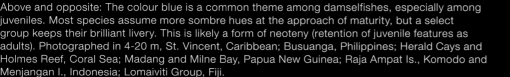

Above and opposite: The colour blue is a common theme among damselfishes, especially among juveniles. Most species assume more sombre hues at the approach of maturity, but a select group keeps their brilliant livery. This is likely a form of neoteny (retention of juvenile features as adults). Photographed in 4-20 m, St. Vincent, Caribbean; Busuanga, Philippines; Herald Cays and Holmes Reef, Coral Sea; Madang and Milne Bay, Papua New Guinea; Raja Ampat Is., Komodo and Menjangan I., Indonesia; Lomaiviti Group, Fiji.

Princess Damsel, *Pomacentrus vaiuli* (3cm)

Springer's Demoiselle, *Chrysiptera springeri* (4 cm)

Neon Damsel, *Pomacentrus coelestis* (4 cm)

Azure Demoiselle, *Chrysiptera hemicyanea* (4 cm)

Optimum coral reef development is correlated with clean, clear water and relatively shallow depths.
Reefs derive their primary energy from solar radiation and thrive in areas of maximum sunlight
where sea temperatures are warm.

Above: Elford Reef, Great Barrier Reef, Australia.
Opposite: Holmes Reef, Coral Sea.

The butterflyfishes and angelfishes are some of the most conspicuous members of the coral reef fish community. Although closely related, the two groups exhibit different mating systems. Most butterflyfishes form long-term pair bonds, while angelfishes typically form harems.

Opposite top: Red Sea Bannerfish, *Heniochus intermedius* (15 cm),
Opposite bottom: Masked Butterflyfish, *Chaetodon semilarvatus* (12 cm), in 10-15 m, Sharm-el-Sheikh, Egypt.
Above: Yellow-masked Angelfish, *Pomacanthus xanthometopon* (32 cm), in 12 m, Tulamben, Bali, Indonesia.

When male mantis shrimps fight, they curl the tail forward exposing bright colour, and use it to block blows delivered by their rivals. Eyespots on the flared tails of crustaceans may also be used to scare potential enemies.

Top and bottom: Rainbow Mantis, *Odontodactylus scyllarus* (7-8 cm), in 10-12 m, Milne Bay, Papua New Guinea.
Opposite Top: Striped Mantis, *Lysiosquillina maculata* (12 cm), in 8 m, Secret Bay, Bali, Indonesia.
Opposite bottom: Penaeid prawn, unidentified (3 cm), in 10 m, Secret Bay, Bali, Indonesia.

Mantis shrimps are classed as either thumpers or spearers depending on the design of their raptorial appendages. When they strike, these limbs are thrown forward with incredible speed and can move at a velocity of over nine metres per second.

Opposite: Rainbow Mantis, *Odontodactylus scyllarus* (15 cm), in 6 m, Milne Bay, Papua New Guinea.
Above: Giant Mantis, *Lysiosquillina lisa* (35 cm), in 10 m, Tulamben, Bali, Indonesia.

Opposite: Seastars have amazing regenerative powers. Eaten by certain fishes, it is not unusual to see even single arms regrowing into complete animals. These photos show one that has regained its outer arms and the other before regeneration commences.

Comb Seastar, *Astropecten polyacanthus* (10 cm), in 8 m, Secret Bay, Bali, Indonesia.

Above: *Tubastrea* corals are commonly encountered in shady positions. Unlike the majority of hard corals, they lack zooxanthellae, the tiny microalgae found in the tissue. This photo provides a rare glimpse of the red mouth, which is usually retracted or hidden by feeding tentacles.

Turret Coral, *Tubastrea faulkneri* (3 cm), in 2 m, Lembeh Strait, Sulawesi, Indonesia.

Living among the stinging tentacles of a sea anemone may seem like a precarious existence, but the anemonefishes have evolved ways to prevent the stinging cells of their hosts from discharging. They develop a chemical camouflage that enables them to live among the tentacles without being detected.

Above: Tomato Anemonefish, *Amphiprion ephippium* (7 cm), in 10 m, Weh Is. Indonesia.
Opposite top: Clown Anemonefish, *Amphiprion percula* (5 cm), in 5 m, Milne Bay, Papua New Guinea.
Opposite bottom: Pink Anemonefish, *Amphiprion perideraion* (6 cm), in 8 m, Komodo, Indonesia.

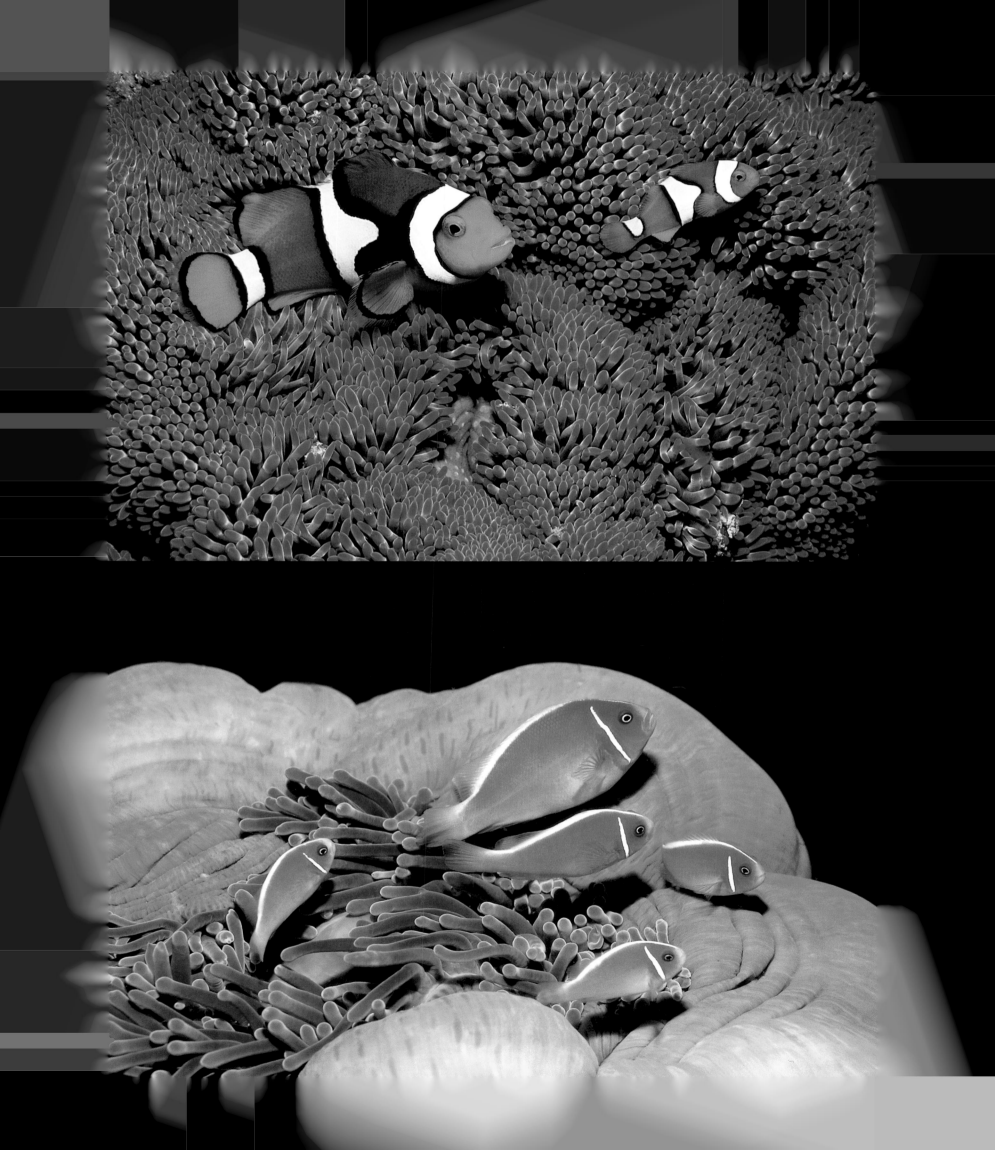

Opposite: There are a number of cardinalfishes, including the Banggai cardinal, which will associate with sea anemones. Unlike the anemonefishes, the apogonids are mildly stung when they contact the tentacles and as a result avoid those sea anemones with a more potent sting.

Banggai Cardinalfish, *Pterapogon kauderni* (4-8 cm), in 10 m, Lembeh Strait, Sulawesi, Indonesia.

Above: Many of the wrasses are opportunistic predators that turn over rocks and rubble with their mouths to expose hidden prey. Brittle stars are favourite food of debris flipping hogfishes.

Spanish Hogfish, *Bodianus rufus* (35 cm), in 12 m, Bonaire, Caribbean.

The groupers are important predators on the coral reef. They may consume as much as five percent of their total weight in food per day to meet their high caloric requirements.

Blue Maori, *Epinephelus cyanopodus* (25 cm), in 20 m, Milne Bay, Papua New Guinea.

Sea slugs of the genus *Thuridilla* are characterized by a slender body and a pair of prominent antler-like rhinophores. They have a well-developed radula and mouth parts used for piercing algal cells and sucking out the contents.

Lined Sea Slug, *Thuridilla lineolata* (2 cm), in 6 m, Lembeh Strait, Sulawesi, Indonesia.

Generally resembling sponges, ascidians (tunicates) are actually more allied to the vertebrates. They circulate sea water through their bodies to filter out minute prey items. Although not large, some may filter as much as 170 litres of water daily.

Above top: Mixed tunicates, blue, *Rhopalaea* sp.; red, *Didemnum* sp. (12 cm), in 10 m, Tulamben, Bali, Indonesia.
Above bottom: Orange ascidians, *Didemnum mosleyi* (25 cm), in 8 m, Milne Bay, Papua New Guinea.
Opposite top: Yellow ascidian, *Botryllus* sp. (10 cm), in 10 m, Komodo, Indonesia.
Opposite bottom: Gold-line tunicate, *Botryllus* sp. (6 cm), in 8 m, Raja Ampat Is, Indonesia.

Opposite: The squat lobster is actually an anomuran crab. They vary in their lifestyles. Some live commensally on sponges, corals and crinoids. Others seek shelter in reef interstices and still others live a pelagic existence.

Squat Lobster, *Galathea pilosa* (2 cm), in 12 m, Coral Sea, Australia.

Above: Ritualized combat between members of the same species commonly occurs on the coral reef. Crabs will fight over hiding places and potential mates. These melees often result in one of the combatants losing one or more appendages, which are usually regenerated.

Clinging Crab, *Mithrax* sp. (4 cm), in 12 m, St. Vincent, Caribbean.

There are a number of shrimps in the genus *Periclimenes* that spend their adult lives riding around on other invertebrates, including sea stars, feather stars, sea cucumbers, and sea slugs. One of these resourceful little crustaceans was even seen hitching a ride on a Flamboyant Cuttlefish.

Opposite top: Imperial Shrimp, *Periclimenes imperator* (4 cm), in 10 m, Milne Bay, Papua New Guinea.
Opposite bottom: Ambon Shrimp, *Periclimenes amboinensis* (2 cm), in 15 m, Lembeh Strait, Sulawesi, Indonesia.
Above: Soror Shrimp, *Periclimenes soror* (1.5 cm), in 15 m, Milne Bay, Papua New Guinea

This Vermetid Snail is a univalve mollusc that settles onto a suitable substrate in larval form to build a protective spiral shell. It feeds by extending mucous strands from within to snare plankton. The colour of the foot is variable.

Vermetid Snail, *Serpulorbis grandis* (2 cm), in 10 m, Milne Bay, Papua New Guinea.

This tropical isopod with its unusual triangular eyes and coat of armour, feeds on the coralline alga in the background. Other species are ferocious carnivores known to reduce carcasses to skin and bone in quick time. Yet others are called sea lice and bite bathers and uncovered divers.

Isopod, unidentified (2cm), in 20 m, Milne Bay, Papua New Guinea.

Overleaf: Intricate details of colour patterns are easily overlooked when an entire organism is viewed. The close-up lens reveals a secret and stunning realm of textures and designs that rival the finest tapestries.

Edithburgh and Port Philip Bay, Australia; New Britain and Milne Bay, Papua New Guinea; Busuanga, Philippines; Secret Bay, Lembeh Strait, Komodo, Indonesia.

Radiant Urchin, *Astropyga radiata*

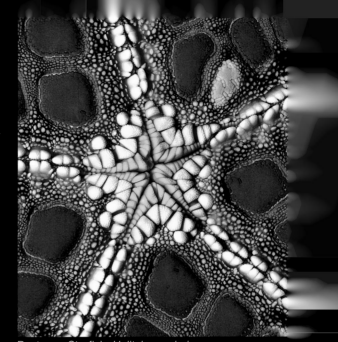

Mushroom Coral, *Fungia* sp.

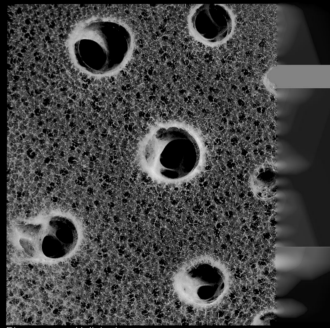

Pentagon Starfish, *Halityle regularis*

Southern Sea Anemone, *Phlyctenanthus australis*

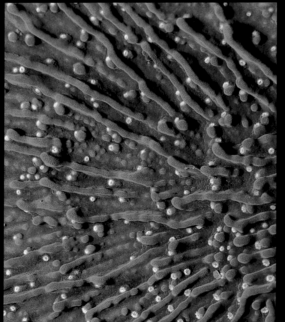

Hard coral, *Montipora foliosa*

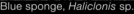

Blue sponge, *Haliclonis* sp.

ther starfish, unidentified

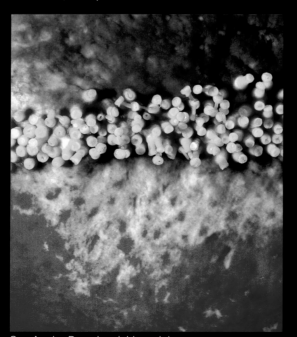

Sea Apple, *Pseudocolochirus violaceus*

Many-armed Starfish, *Coscinasterias muricata*

Starfish, *Echinaster callosus*

Sea urchin, *Diadema* sp.

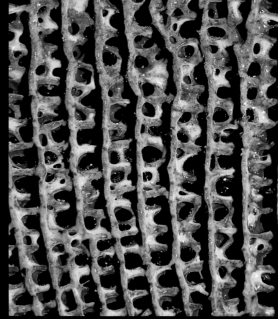

Elephant-ear Sponge, *Ianthella basta*

Upside-down Jellyfish, *Cassiopea andromeda*

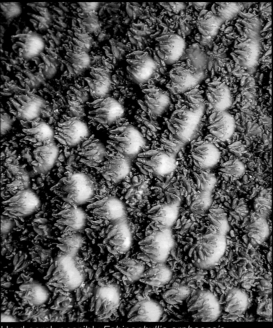

Hard coral, possibly *Echinophyllia orpheensis*

Red-lined Sea Cucumber, *Thelenota rubrolineata*

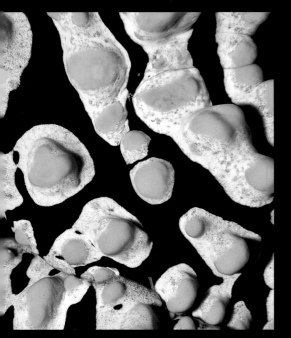

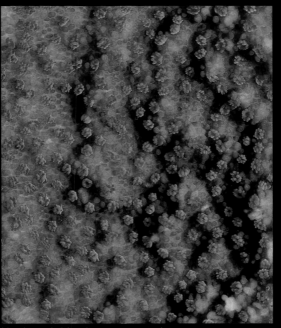

Many of the facelinid nudibranchs feed on stinging hydroids. The undigested stinging cells of their prey are transported into the cerata on their backs and are then used as a form of defence.

Indian Slug, *Phidiana indica* (4 cm), in 15 m, Milne Bay, Papua New Guinea.

The gobies are the largest family of reef fishes. Many have evolved to occupy specific niches. This orange convict goby is a secretive species that often hangs upside down in caves, crevices and under benthic debris.

Orange convict goby, *Priolepis* sp. (3 cm), in 20 m, Lembeh Strait, Sulawesi, Indonesia.

Ribbon Sweetlips, *Plectorhinchus polytaenia*

Raggy Scorpionfish, *Scorpaenopsis venosa*

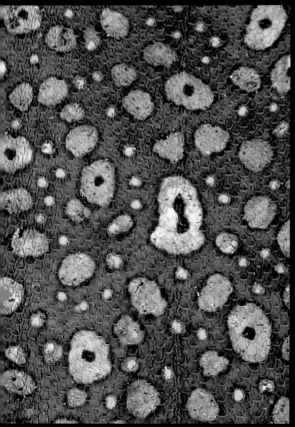

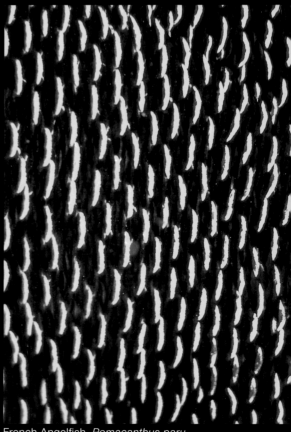

Peacock Sole, *Pardachirus pavoninus*

French Angelfish, *Pomacanthus paru*

Fashion designers need look no further than the inspiring patterns offered by nature's exquisite creatures. Fishes are a perfect example of the boundless possibilities. Fins, skin, and scales are marked with an endless variety of combinations.

Tulamben and Lembeh Strait, Indonesia; Milne Bay, Papua New Guinea and Bonaire, Caribbean.

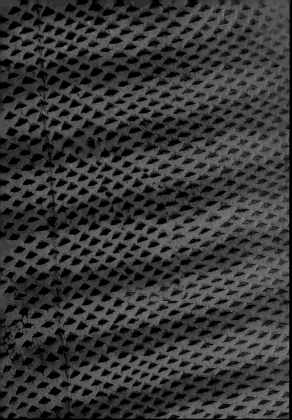

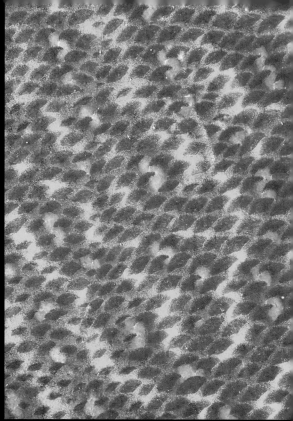

Ring Wrasse, *Hologymnosus doliatus*

Lyretail Grouper, *Variola albomarginata*

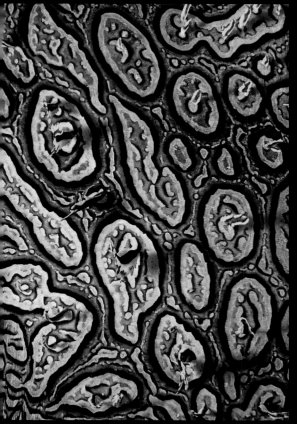

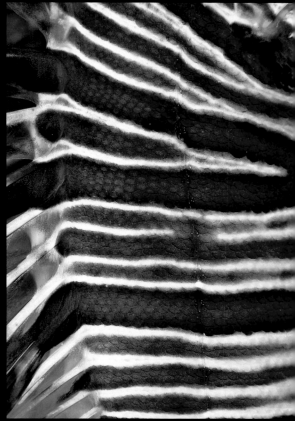

Echinoderms are one of the most variable groups of marine organisms. Their tremendous diversity in external appearances belies a common internal structure based on radial symmetry.

Opposite: Sea urchin, *Diadema* sp. (30cm), in 10m, Secret Bay, Bali, Indonesia.
Above: Ludwig's Basket Star, *Conocladus ludwigi* (12cm), in 12m, Milne Bay, Papua New Guinea.
Overleaf left: Feather starfish, unidentified comasterid (12 cm), in 12m, Milne Bay, Papua New Guinea
Overleaf right: Starfish, *Pentaster obtusatus* (field of view 5cm), in 15m, Lembeh Strait, Sulawesi, Indonesia.

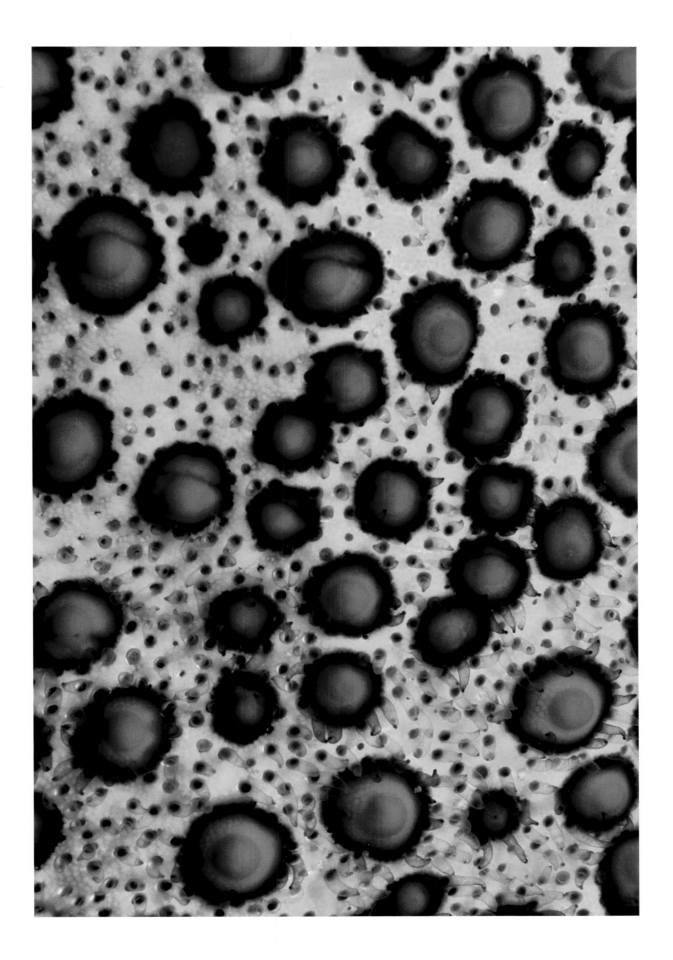

Above: An endless variety of islands, large and small, are a paramount feature of the world's tropical seas. Cays and rocky islets are especially interesting for underwater exploration as they offer excellent ecological and biological diversity.

Amphlette Is., Milne Bay, Papua New Guinea.

Opposite: The seldom visited Raja Ampat Islands off western New Guinea, have suddenly emerged as a diving hotspot. A recent marine survey by Conservation International revealed the highest number of hard corals ever recorded in one area and the most fish species seen during a single dive.

Raja Ampat Is., Indonesia.

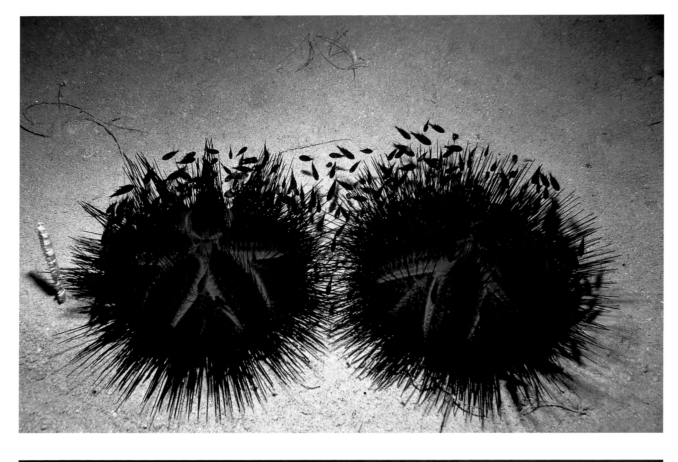

Above and opposite: These photos document the cooperative hunting in a "pride" of lionfishes. The lionfish surrounded urchins burgeoning with cardinalfish, and brushed their long pectoral rays into the urchin, startling the small fish. This caused them to temporarily abandon the safety of the urchin's spines and swim into striking range of neighbouring lionfishes.

Lionfishes, *Pterois volitans* and *Pterois russelli* (15-20 cm),
Radiant Urchin, *Astropyga radiata*, in 12 m, Milne Bay, Papua New Guinea.

Above and opposite: Undoubtedly the Queen of nudibranchs, the Spanish Dancer derives its common name from the flared skirt effect produced by the undulations it displays while swimming. Widespread in the Indo-Pacific, it has variable colouration and can attain a size of 50 cm.

Spanish Dancer, *Hexabranchus sanguineus* (40cm), in 5 m, Milne Bay, Papua New Guinea.

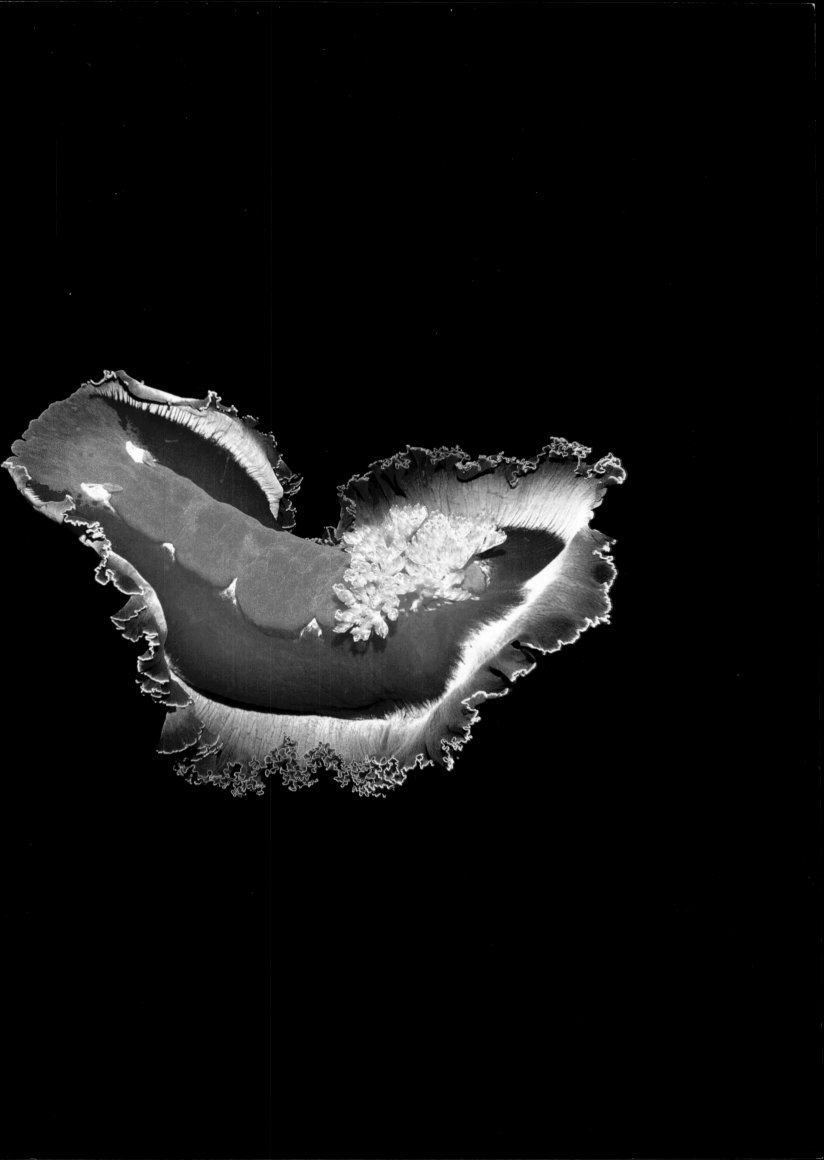

The pycnogonids or sea spiders are a diverse group numbering approximately 1500 species that are found in many different habitats, including deep oceans and under the polar ice. There are also species on temperate rocky reefs and tropical coral reefs. Those from the former tend to be more colourful than their warm water counterparts.

Above: Tropical sea spider, *Nymphosus* sp. (2 cm), in 15 m, Lembeh Strait, Sulawesi, Indonesia.
Opposite: Temperate sea spider, *Anoplodactylus evansi* (2 cm), in 6 m, Hopetoun, Western Australia.
(Photo C. Bryce)

Although armed with deadly neurotoxic venom, the blue-ringed octopuses are responsible for few fatalities. Most of the bites that have been reported occurred when the animal was being handled or provoked in some way. The venom is produced in the salivary gland and used to paralyse the crabs they feed on.

Above top: Mid-ring Blue-ringed Octopus, *Hapalochlaena* sp. (8 cm), in 20 m, Lembeh Strait, Sulawesi, Indonesia.
Above bottom: Southern Blue-ringed Octopus, *Hapalochlaena maculosa* (8 cm), in 5 m, Edithburgh, South Australia.
Opposite top: Greater Blue-ringed Octopus, *Hapalochlaena lunulata* (6 cm), in 1 m, Tukang Besi, Indonesia.
Opposite bottom: Southern Blue-ringed Octopus, *Hapalochlaena maculosa* (8 cm), in 10 m, Port Phillip Bay, Victoria, Australia.

Resembling amphibious tanks, many shells erupt from the substrate at night to hunt their quarry. The large siphon visible on some species is used to take in water to irrigate the gills and also acts as a chemoreceptor.

Above Top: Undulate Moon Snail, *Tanea undulata* (7 cm).
Above bottom: Bonnet Shell, *Semicassis bisulcata* (5 cm).
Opposite top: Baler Shell, *Melo amphora* (8 cm), in 10-12 m, Milne Bay, Papua New Guinea.
Opposite bottom: Harp Snail, *Harpa articularis* (7 cm), in 15 m, Lembeh Strait, Sulawesi, Indonesia.

Above and opposite: The colourful pigments seen on reef fishes serve a variety of functions. One important purpose is to enable species to identify members of their own kind. The ornate markings on this male Choat's Wrasse are like an identification card that signals others of its presence.

Choat's Wrasse, *Macropharyngodon choati* (8 cm), in 2 m, Great Barrier Reef, Australia.

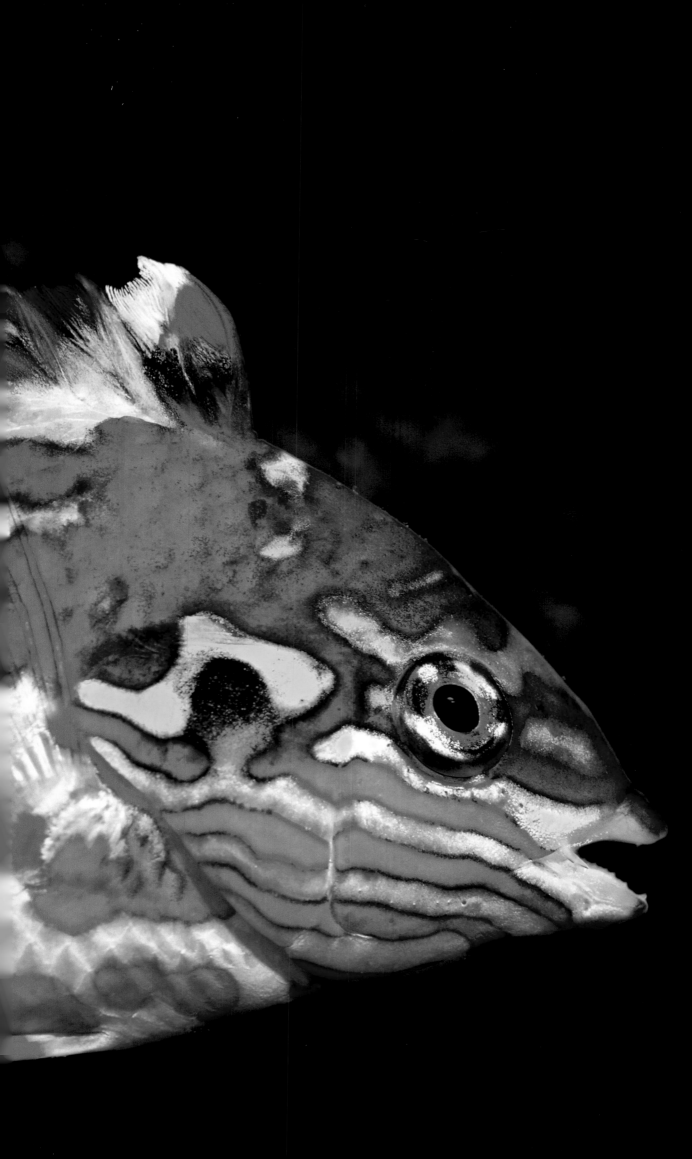

Bluestripe Fairy Wrasse, *Cirrhilabrus temmincki* (9 cm)

Flame Wrasse, *Cirrhilabrus jordani* (8 cm), (photo F. Walsh)

Red-Margin Wrasse, *Cirrhilabrus rubrimarginata* (9 cm)

Fiji Fairy Wrasse, *Cirrhilabrus* sp. (6 cm)

Spotted Fairy Wrasse, *Cirrhilabrus punctatus* (10 cm)

Coral Sea Fairy Wrasse, *Cirrhilabrus bathyphilus* (5 cm), (F. Walsh)

Filamented Flasher Wrasse, *Paracheilinus filamentosus* (6 cm)

Spanish Hogfish, *Bodianus rufus* (30 cm)

Containing approximately 500 species, the wrasse family is the second most speciose fish group in the world. They include some of the most colourful fishes found in the oceans, with almost every hue of the colour spectrum being represented.

In 8- 35 m, Bonaire, Florida, Hawaii, Fiji, Japan, Coral Sea, Great Barrier Reef, Western Australia, Indonesia and Egypt.

Lennard's Wrasse, *Anampses lennardi* (15 cm), (photo F. Walsh)

Eightline Flasher Wrasse, *Paracheilinus octotaenia* (8 cm)

Rooster Hogfish, *Lachnolaimus maximus* (30 cm)

Torpedo Wrasse, *Pseudocoris heteroptera* (12 cm), (photo F. Walsh)

Marjorie's Fairy Wrasse, *Cirrhilabrus marjorie* (7 cm)

Lined Fairy Wrasse, *Cirrhilabrus lineatus* (8 cm)

Pacific Exquisite Wrasse, *Cirrhilabrus* sp. (6 cm)

Laboute's Fairy Wrasse, *Cirrhilabrus laboutei* (8 cm)

There are a number of shrimps that make a living cleaning fishes. As they groom their clients, these crustaceans ingest small parasites, injured or dead tissue, scales and body slime. Morays are a favourite customer of many cleaner shrimps.

Above: Scarlet Cleaner Shrimp, *Lysmata debelius* (5cm), Giant Moray, *Gymnothorax javanicus*, in 18 m, Lembeh Strait, Sulawesi, Indonesia.
Opposite: Ambon Cleaner Shrimp, *Lysmata amboinensis* (5 cm), Giant Moray, *Gymnothorax javanicus*, in 15 m, Komodo, Indonesia.

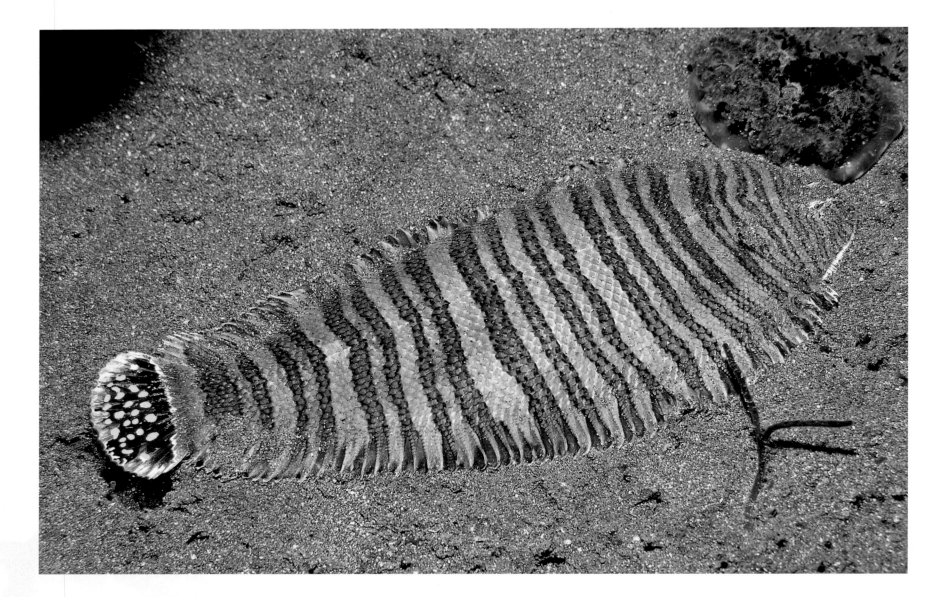

Some flatfishes use their fins to communicate to each other and potential predators. The Zebra Sole has a tail that looks like a noxious flat worm that may discourage fish attacking. The Cockatoo Flounder has white, elongated dorsal rays that are thrown out and forward when it is threatened or when it encounters a member of its own kind.

Above: Zebra Sole, *Zebrias zebra* (20 cm), in 10 m, Secret Bay, Bali, Indonesia.
Opposite: Cockatoo Flounder, *Samaris cristatus* (10 cm), in 20 m, Lembeh Strait, Sulawesi, Indonesia.

Opposite: In recent years, it has been discovered that at least one member of the anemone-like corallimorpharians can catch and consume certain echinoderms. This species uses its sticky tentacles to capture sea stars, urchins, holothurians and feather stars.

Corallimorpharian, *Pseudocorynactis* sp. (30 cm), in 6 m, Lembeh Strait, Sulawesi, Indonesia.

Above: Jacks often follow other fishes, including those that disturb the sea floor and expose hidden prey that they can then pounce on. They also use larger fishes as a moving blind, and swim close to their associate until they are close enough to launch an attack at unsuspecting prey.

Fine-spotted Porcupinefish, *Diodon holocanthus* (30 cm), Barred Jack, *Caranx* sp., in 20 m, Lembeh Strait, Sulawesi, Indonesia.

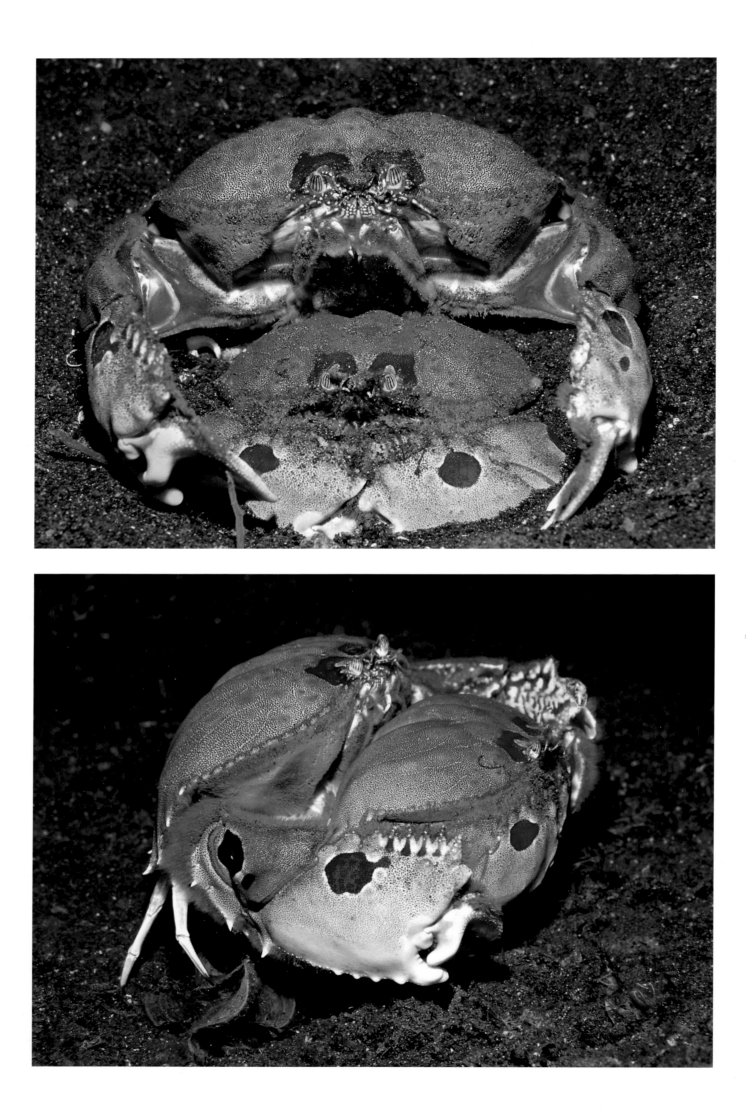

The male box crab coaxes the female out of her sandy sanctuary, then carries her off to mate. When eggs are fully developed, gravid females release the hatched zoea into the water column where they undergo a series of growth stages before settling permanently onto the substrate.

Opposite: Box crabs, *Calappa philargius* (11cm), in 5 m, Lembeh Strait, Sulawesi, Indonesia.
Above: *Calappa lophos* (10 cm), in 10 m, Osezaki, Japan. (photo S. Michael)

Sponge crabs employ a unique camouflage strategy. They find an appropriate sponge or tunicate, remove it, and shape it with their claws. They then use their rear appendages to hold it on top of their domed carpace like a fancy hat. The beret continues to live and grow.

Above top: Sponge crabs, *Cryptodromia octodentata* (8 cm), in 6 m, Edithburgh, South Australia.
Above bottom: *Dromidopsis dubia* (3 cm), in 12 m, Lembeh Strait, Sulawesi, Indonesia.
Opposite top: *Lauridromia* sp. (2 cm), in 10 m, Lembeh Strait, Sulawesi, Indonesia.
Opposite bottom:*Cryptodromia* cf. *coronata* (2 cm), in 12 m, Milne Bay, Papua New Guinea.

The travelling Carrier Crab utilizes rear legs to hold objects such as venomous sea urchins, jellyfish and sticks on top of its body to prevent predatory attack. In addition to the Toxic Urchin, the individual above also travels with a female slung underneath.

Above and opposite: Carrier Crab, *Dorippe frascone* (10-15 cm), in 6-8 m, Lembeh Strait, Sulawesi, Indonesia.

Overleaf: Close inspection of the reef's surface reveals exotic miniature gardens of sessile invertebrates. Sponges, ascidians, corals and algae are the underwater equivalent of wildflowers and combine to form brilliant bouquets.

Mixed invertebrates, unidentified, Milne Bay, Papua New Guinea.

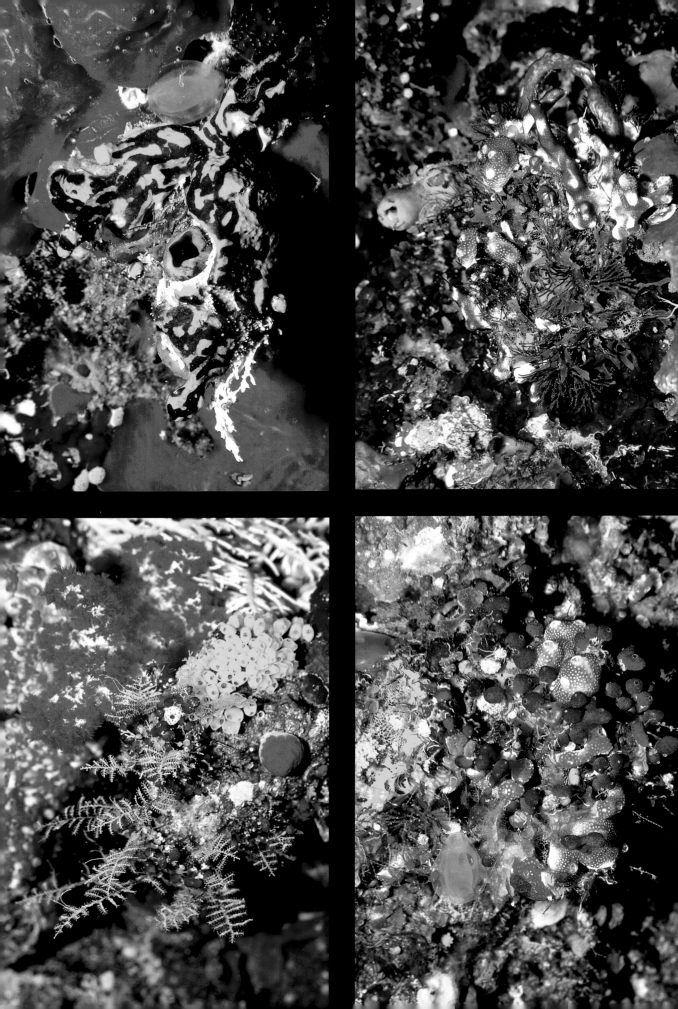

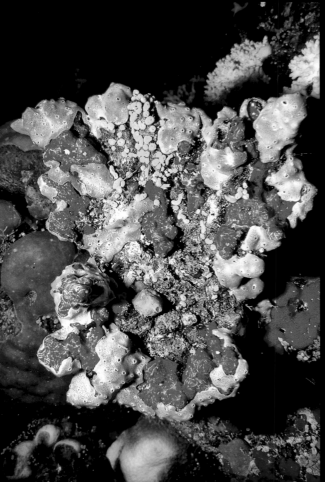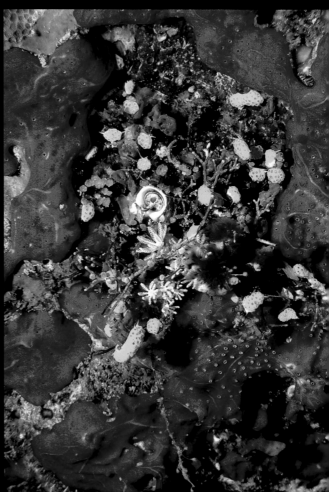

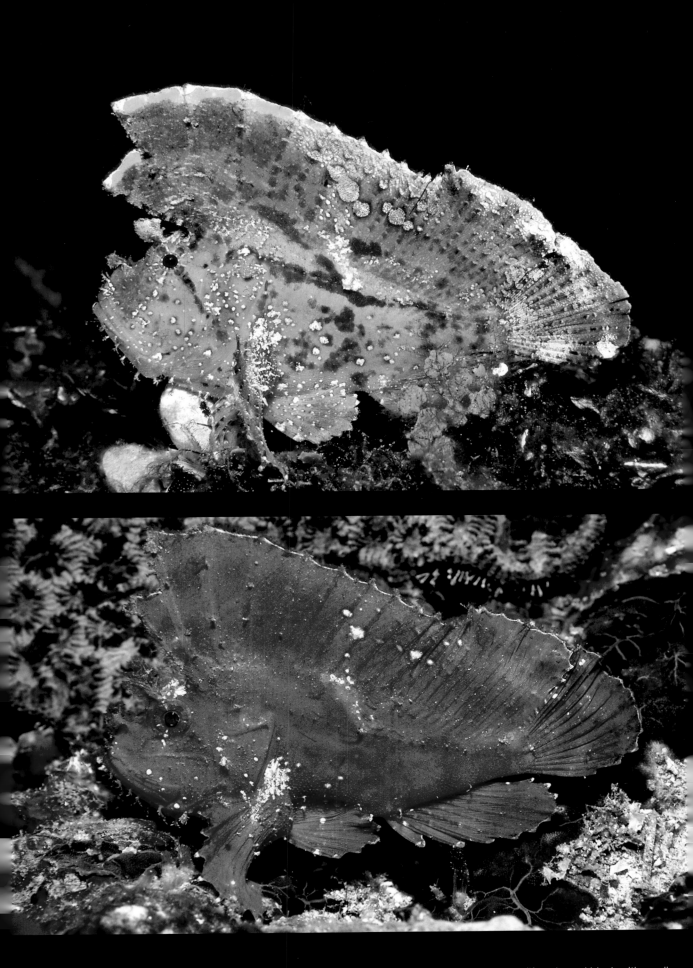

Opposite: One of the most elegant animals to be seen near coral reefs is the ribbon eel. The young start life coloured black with a yellow dorsal fin but change into the blue male form as they mature. Large males ultimately develop female sex organs and change colour to yellow

Blue Ribbon Eel, *Rhinomuraena quaesita* (80 cm), in 10 m, Lembeh Strait, Sulawesi, Indonesia.

Above: The Leaf Scorpionfish not only looks like a piece of plant material, it facilitates this resemblance by rocking from side-to-side. This makes it look even more like a piece of macro-algae or a waterlogged leaf being buffeted by the surge.

Leaf Scorpionfish, *Taenianotus triacanthus* (7-8 cm), in 12-15 m, Lembeh Strait, Sulawesi, Indonesia.

The ophiuroids employ a number of different antipredation strategies. Some brittle stars coil tightly around sea fan branches or corals during the day and unfurl their arms at night to snag drifting particles. Others hide under rubble or in reef crevices and come out to feed after dark.

Above: Dana's Brittle star, *Ophiothela danae* (2 cm), in 8 m, Komodo, Indonesia.
Opposite top: Sea rod Basket star, *Schizostella bifurcata* (1.5 cm), in 12 m, Cozumel, Mexico.
Opposite bottom: Brittle Star, Ophiuroidea (6 cm), in 15 m, Busuanga, Philippines.

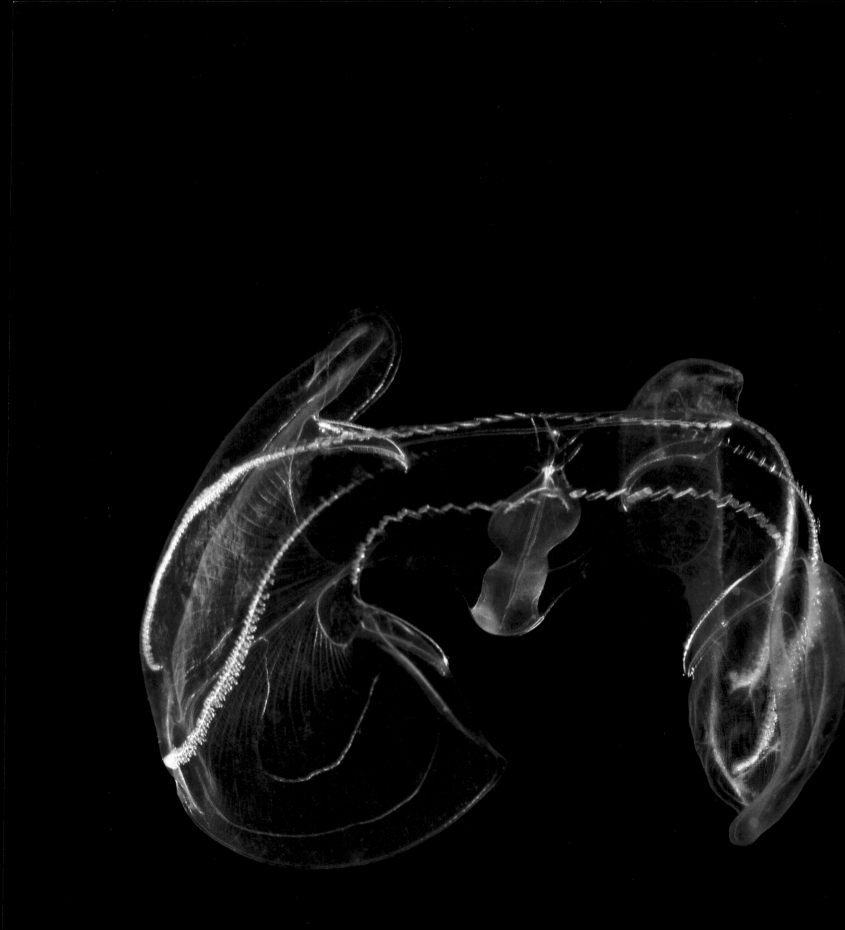

Comb jellies inhabit the open sea, but often drift into the shallows. The Winged Comb Jelly propels itself with an exaggerated flapping motion of the oral lobes. When disturbed at night it emits a greenish-blue bioluminescence.

Winged Comb Jelly, *Ocyropsis crystallina* (7 cm), in 2 m, Great Barrier Reef, Australia.

Miniscule corycaesid copepods are planktonic predators with pronounced bifocal vision. The unusual eye structure consists of a retina that protrudes from the bottom of the body while the remainder of the eye is directed upwards.

Corycaesid copepod, Great Barrier Reef, Australia.

The dragonets occur in temperate and tropical seas. While many exhibit more muted colours that help them blend in with the substrate, the family also includes some of the most ornately attired fishes in the ocean.

Above top: Psychedelic Mandarinfish, *Synchiropus picturatus* (5 cm), in 8 m, Secret Bay, Bali, Indonesia.
Above Bottom: Splendid Mandarinfish, *Synchiropus splendidus* (4 cm), in 2 m, Palau.
Opposite top: Painted Stinkfish, *Eocallionymus papilio* (10 cm), in 10 m, Port Philip Bay, Victoria, Australia.
Opposite bottom: Moyer's Dragonet, *Synchiropus moyeri* (5 cm), in 9 m, Tulamben, Bali, Indonesia.

The pipefishes are close relatives of the seahorses. Many depend on blending in with their surroundings to avoid being eaten. Other than the bony rings around the body and skin appendages that assist with camouflage, they have no defences.

Top and bottom: Winged Pipefish, *Halicampus macrorhynchus* (8-12 cm), in 8-15 m, Milne Bay, Papua New Guinea.

There are other pipefishes that are brightly coloured and hover above the bottom in caves and crevices. Some of these species are known to clean cryptic fishes, like morays and cardinalfishes. Their bold colour patterns may advertise their cleaning services.

Top: Japanese Pipefish, *Doryrhamphus japonicus* (6 cm),
Bottom: Yellow-Banded Pipefish, *Dunkerocampus pessuliferus* (10 cm), in 12-20 m, Lembeh Strait, Sulawesi, Indonesia.

The sea dragons are the supreme masters of concealment. Although they may be easy to observe when away from their preferred habitat, they become inconspicuous against plant fronds when among rocky reefs with kelp and sargassum weed.

Above: Leafy Sea Dragon, *Phycodurus eques* (28 cm), in 8 m, Edithburgh, South Australia.
Opposite: Weedy Sea Dragon, *Phyllopteryx taeniolatus* (32 cm), in 10 m, Port Philip Bay, Victoria, Australia.

Above and opposite: Going! Going! Gone! When in the open water, the Halimeda Ghost Pipefish is very conspicuous. But as it moves near the calcareous algae that it resembles, its remarkable adaptation is truly obvious.

Halimeda Ghost Pipefish, *Solenostomus halimeda* (7 cm), in 12 m, Madang, Papua New Guinea.

Above and opposite: The beauty of coral reefs in the Caribbean rivals those in the Indo-Pacific. Sponges figure prominently in the aquascape on the reefs off the Yucatan peninsula. The water is usually gin clear in this region, because of the lack of heavy rainfall during most times of the year.

Underwater gardens, mixed invertebrates, unidentified, in 20 m, Cozumel, Mexico.

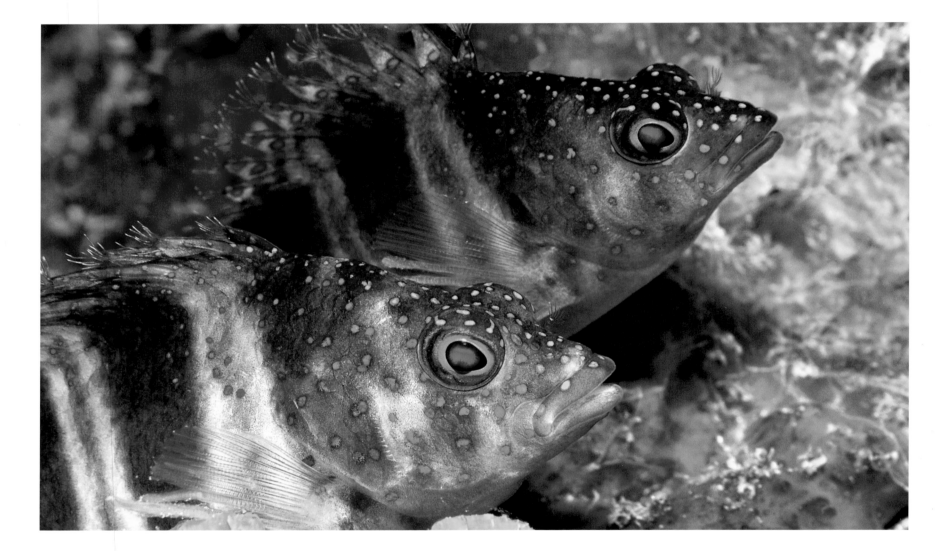

Above: The hawkfishes spend most of the time perched on the substrate. They are characterized by the presence of cirri (hair-like filaments) on their single dorsal fin and behind the nostrils.

Redspotted Hawkfish, *Amblycirrhitus pinos* (6 cm), in 12 m, Bonaire, Caribbean.

Opposite: A number of nocturnal fishes, like these squirrelfishes, are red in colour. Because red is filtered out at shallow depths, they look grey or black at depth. Being brightly coloured makes no difference at night, when there is a minimal amount of light present.

Opposite top: Three-spot Squirrelfish, *Sargocentron melanospilos* (18 cm), in 12 m, Lembeh Strait, Sulawesi, Indonesia.
Opposite bottom: Longjaw Squirrelfish, *Sargocentron marianus* (12 cm), in 20 m, St. Vincent, Caribbean.

Not only humans like eating shrimps, they are also a favourite with a myriad of undersea animals. Some fishes feed on crustaceans during the day by digging them out of the sand or coral rubble or by plucking them out of reef interstices with modified jaws.

Above top: Highfin Perchlet, *Plectranthias* sp. (5 cm), in 20 m, Milne Bay, Papua New Guinea.
Above bottom: Weedfish, *Heteroclinus tristis* (12 cm), in 10 m, Port Philip Bay, Victoria, Australia.
Opposite top: Peacock Razorfish, *Xyrichtys pavo* (8 cm), in 10 m, Lembeh Strait, Sulawesi, Indonesia.
Opposite bottom: Coral Cod, *Cephalopholis miniata* (8 cm), in 10 m, Maumere, Flores, Indonesia.

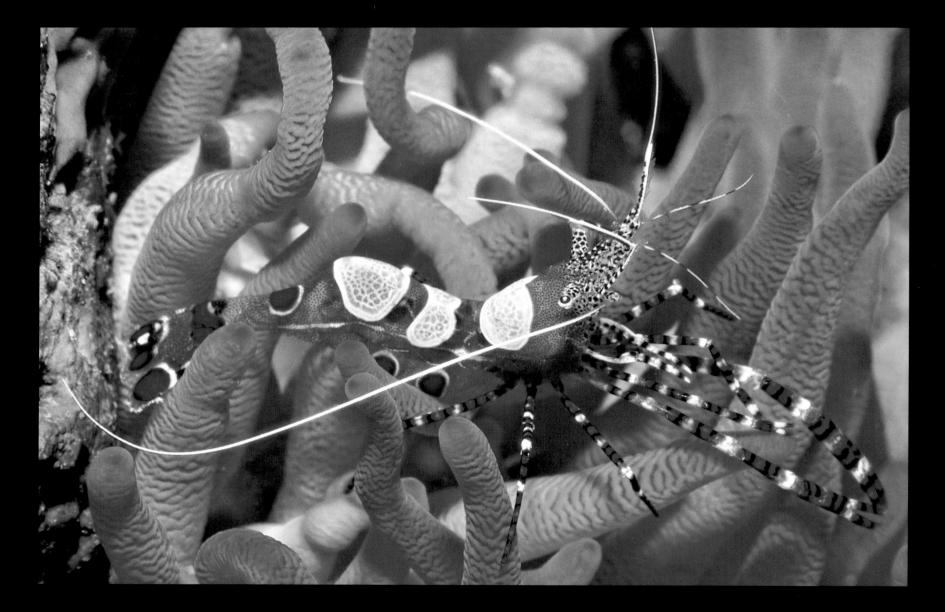

The beautiful Spotted Cleaner Shrimp is dependent on its sea anemone host to avoid being eaten by predators. It is also a cleaner that will remove necrotic tissue and parasites from fishes that pose near its cnidarian home.

Spotted Cleaner Shrimp, *Periclimenes yucatanicus* (4 cm), in 10 m, Bonaire, Caribbean.

Paguristes cadenati (4 cm)

Calcinus minutus (2 cm)

Pagurus sp. (3 cm)

Calcinus morgani (1.5 cm)

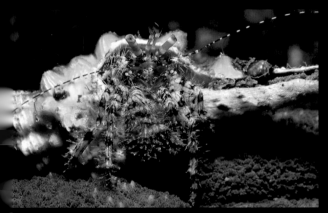

Dardanus sp. (3 cm)

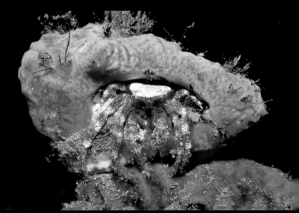

Paguristes sp. (4 cm)

Dardanus guttatus (7 cm)

Ciliopagurus strigatus (5 cm)

Pagurus similis (5 cm)

Aniculus miyakei (6 cm)

Calcinus elegans (2 cm), (photo C. Bryce)

Dardanus crassimanus (6 cm)

Pylopaguropsis zebra (2 cm)

Clibanarius virescens (3 cm)

Some fishes will develop a specific search image and feed heavily on a single type of food. Large Map Puffers sometimes gorge on sponges or detritus until their bellies are distended and they are unable to lift off the bottom to swim.

Map Puffer, *Arothron mappa* (60 cm), in 10 m, Milne Bay, Papua New Guinea.

The walking batfishes or pancake fishes are true piscine oddities that are found on deep sand or mud slopes. However there are a few species that are encountered at safe diving depths. These odd creatures have leg-like paired fins that are used to "walk" over the sea floor.

Walking Batfish, *Halieutaea indica* (10 cm), in 10 m, Secret Bay, Bali, Indonesia.

Tropical and polar seas offer vivid contrasts. Contrary to popular belief, equatorial oceans are notoriously poor in nutrients, but solar energy is efficiently converted by shallow reef ecosytems to support a wealth of biodiversity. Polar seas, although rich in nutrients, are inhabited by relatively few organisms.

Opposite: Aerial, Palau.
Above: Aerial, Antarctica.

There are a number of sea horses found in temperate waters. These species often have more skin ornamentation that helps them blend in with the backgrounds found in cooler seas.

Above: Pot-belly Seahorse, *Hippocampus bleekeri* (25 cm), in 8 m,
Port Philip Bay, Victoria, Australia.
Opposite: Painted Seahorse, *Hippocampus sindonis* (4 cm), in 8 m, Oshima I., Japan.

The 4,500 species of true crabs exhibit great diversity in size, colour and behaviour. The leg span of the Japanese spider crab can be over one metre in width, while the pea crab is as small as its vegetable namesake.

Opposite: Squat Lobster, *Galathea* sp. (2 cm), in 12 m, Coral Sea, Australia.
Above: Porcelain Crab, *Petrolisthes coccineus* (3 cm), in 15 m, Tulamben, Bali, Indonesia.

Like a ball of serpents, a Caribbean basket star wraps around a soft coral during the day (opposite). At night, it will unfurl and feed. It is undetermined if the eggs enveloping the Indo-Pacific basket star (above) came from the echinoderm or the underlying coral.

Opposite: Giant Basket Star, *Astrophyton muricatum* (15 cm), in 10 m, St. Vincent, Caribbean.
Above: Basket star, *Astroboa nuda* (40 cm), in 8 m, Tulamben, Bali, Indonesia.

The gobies (see also overleaf), comprise the largest family of fishes with over 2000 species. Some, like the shrimp gobies, even form partnerships and shelter with unrelated marine organisms, behaviour known as interphyletic symbiosis. Photographed in 1-35 m, Milne Bay, Papua New Guinea; Secret Bay, Tulamben, Lembeh Strait, Indonesia and Osezaki, Japan.

Above: Flag-tail Shrimp Goby, *Amblyeleotris yanoi* (8 cm), Randall's Shrimp, *Alpheus randalli*, in 6 m, Tulamben, Bali, Indonesia.
Opposite: Lined Dartfish, *Ptereleotris grammica* (6 cm), in 50 m, Menjangan I., Bali, Indonesia.

Spotted Shrimp Goby, *Amblyeleotris guttata* (6 cm)

Pink Shrimp Goby, *Cryptocentrus leptocephalus* (8 cm)

Candycane Dwarf Goby, *Trimma* sp. (2 cm)

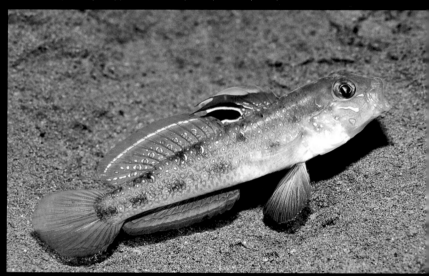

Spinecheek Goby, *Oplopomus oplopomus* (7 cm)

Giant Shrimp Goby, *Amblyeleotris fontanesii* (15 cm)

Metallic Shrimp Goby, *Amblyeleotris latifasciata* (8 cm)

Steinitz' Shrimp Goby, *Amblyeleotris steinitzi* (6 cm)

Wheeler's Shrimp Goby, *Amblyeleotris wheeleri* (4 cm)

Orangeline Goby, *Pterogobius virgo* (16 cm)

Ward's Goby, *Valenciennea wardii* (8 cm)

The abundance of *Dendronephthya* has earned Fiji the title of soft coral capital of the world. Although common throughout the Indo-Pacific, there are few locations where it achieves such luxuriant growth.

Lace corals of the family Stylasteridae resemble hard corals but are actually hydrozoans, a group that includes fire corals and stinging hydroids. Most species form brittle, delicate branches, but a few have knobby, blunt tips.

Opposite and above: Lace coral, *Distichopora* spp. (15-35 cm), in 2 m, Raja Ampat Is., Indonesia.
Overleaf left: Delicate Lace Coral, *Stylaster* sp. (15 cm), in 15 m, Ambon, Indonesia.
Overleaf right: Blunt Lace Coral, *Distichopora vervoorti* (15 cm), in 5 m, Komodo, Indonesia.

Above and opposite: The eyes of opistobranchs are either poorly developed or absent, at best serving as simple light receptors as in the Snakey Bornella. This species also has a distinctive, eel-like swimming motion consisting of lateral undulations that facilitates rapid movement.

Snakey Bornella, *Bornella anguilla* (4 cm), in 3 m, Milne Bay, Papua New Guinea.

Until recent years pygmy seahorses were virtually unknown. Over the past decade an increasing number of sightings have been reported by divers in the Indo-Pacific. The latest discoveries in this intriguing group, both new to science, are featured here.

Above: Coleman's Pygmy Seahorse, *Hippocampus colemani* (1 cm), in 10 m,
Opposite: Milne Bay Pygmy Seahorse, *Hippocampus* sp. (5 mm), in 12 m, Milne Bay, Papua New Guinea.

Fishes yawn for varying reasons including being cleaned and stretching. Some scorpionfishes unfold bright colours not normally seen as a warning. Frogfish may yawn as displacement behaviour, releasing nervous energy when a threat is perceived. They can increase the oral cavity 12 times normal volume.

Opposite top: Coral Cod, *Cephalopholis miniata* (25 cm), in 10 m, Milne Bay, Papua Bay Guinea,
Opposite bottom:Papua Scorpionfish, *Scorpaenopsis papuaensis* (20 cm), in 10 m, Loloata I., Papua New Guinea,
Above top: Devil Scorpionfish, *Scorpaenopsis diabola* (28 cm),
Above bottom: Striated Frogfish, *Antennarius striatus* (15 cm), in 5- 15 m, Lembeh Strait, Sulawesi, Indonesia.

Looking much like an eel, the seldom seen Bandfish occurs in colonies, each individual having its own muddy or sandy burrow. When the current starts running, they dart upwards and hover in a vertical posture to snatch passing plankton before reversing back into the hole.

Bandfish, *Acanthocepola abbreviata* (12 cm), in 20 m, Milne Bay, Papua New Guinea.

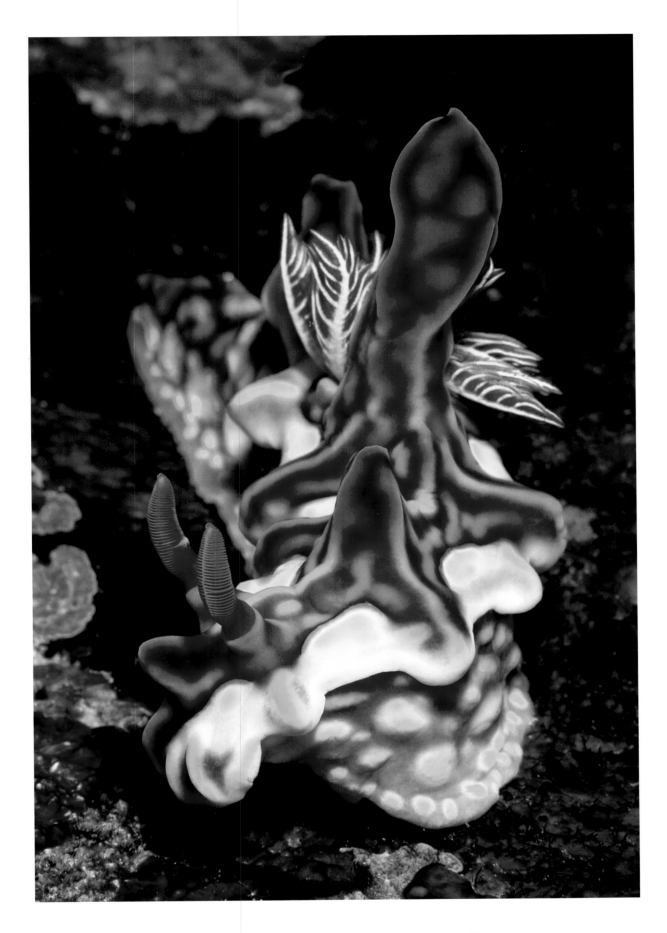

There are approximately 2000 nudibranch species known from the Indo-Pacific. Their brilliant colours, strange shapes and beautiful patterns ensure interest will never wane in this outstanding group. Most occur in tropical waters with many more awaiting discovery.

Magnificent Ceratosoma, *Ceratosoma magnifica* (6 cm), in 18 m, Milne Bay, Papua New Guinea.

The Red Emperor is a highly prized table fish. Juveniles live in sheltered, inshore habitats and seek refuge among sea urchin spines. Adults live deeper and attain a uniform red colouration.

Red Emperor, *Lutjanus sebae* (20 cm), in 12 m, Milne Bay, Papua New Guinea.

Above and overleaf: The ovulid snails or allied cowries are highly specialized gastropods that feed on soft corals, sea fans and black corals. Many of them live commensally on their invertebrate prey. In some species, the protuberances on the mantle closely resemble the expanded polyps of their hosts. Overleaf, in 6-20 m, Ambon, Bali, Flores, Komodo and Lembeh Strait, Indonesia; Milne Bay, Papua New Guinea; Bonaire and St. Vincent, Caribbean.

Above: Rosy Spindle Cowry, *Phenacovolva rosea* (3 cm), in 25 m, Osezaki, Japan.

Minor Spindle Cowry, *Phenacovolva* sp. (3 cm)

Dondan's Egg Cowry, *Serratovolva dondani* (1.5 cm)

Toe-nail Egg Cowry, *Calpurnus verrucosus* (2 cm)

Rosy Spindle Cowry, *Phenacovolva rosea* (3 cm)

Burgundy Spindle Cowry, *Phenacovolva* sp. (3 cm)

Rosewater's Egg Cowry, *Primovula rosewateri* (1 cm)

Milky Egg Cowry, *Calpurnus lacteus* (2 cm)

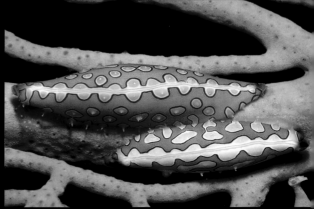

Brown-tip Spindle Cowry, *Phenacovolva brunneiterma* (3 cm)

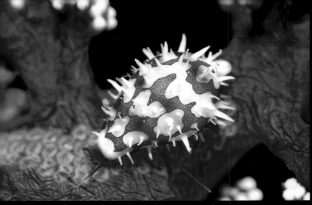

Ridged Egg Cowry, *Pseudosimnia culmen* (1 cm)

Flamingo Tongue Cowry, *Cyphoma gibbosum* (2 cm)

Pirie's Egg Cowry, *Prosimnia piriei* (1.5 cm)

Graceful Spindle Cowry, *Phenacovolva gracilis* (3 cm)

Compressed Spindle Cowry, *Phenacovolva coarctata* (3 cm)

Fingerprint Cowry, *Cyphoma signatum* (2 cm)

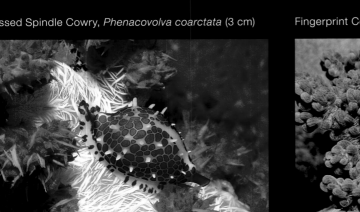

Golden Egg Cowry, *Pseudosimnia* sp. (1 cm)

Margarita Egg Cowry, *Pseudosimnia margarita* (1 cm)

The snorkel apparatus visible on this Banded Sole is a nostril adaptation that allows the fish to respire while buried. There are over 100 species of soles, all live in a sand or soft bottom environment that allows them to bury quickly.

Banded Sole, *Soleichthys heterohinos* (10 cm), in 5 m, Milne Bay, Papua New Guinea.

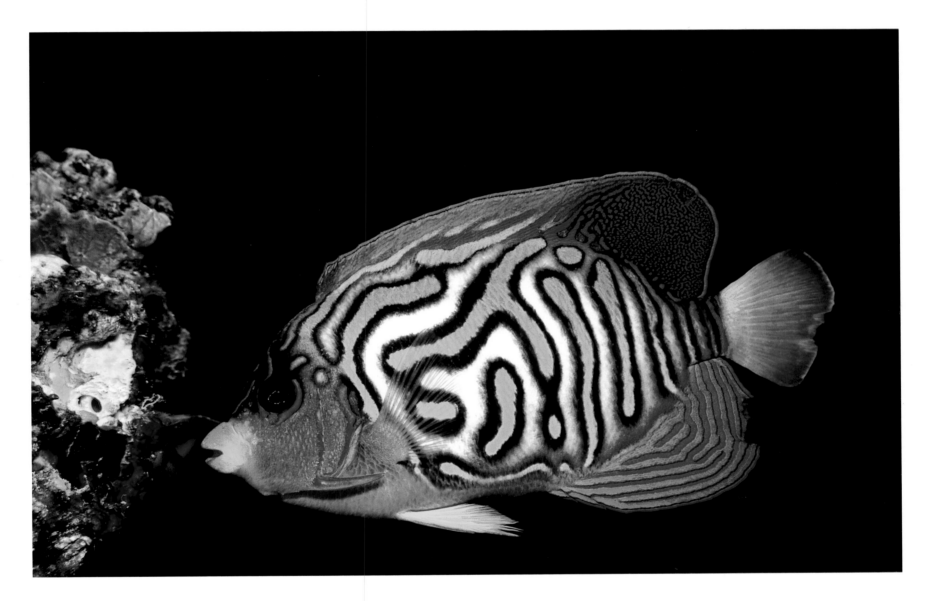

While the colour patterns of angelfish are consistent enough to be a reliable aid in species identification, a rare individual may exhibit aberrant colour characteristics. Both a normal (inset) and unusual form of the Regal Angelfish is shown here.

Regal Angelfish, *Pygoplites diacanthus* (12-20 cm), in 6-8 m, Red Sea, Egypt and Milne Bay, Papua New Guinea.

Above and opposite: The colourful fairy basslets are small, spectacular members of the grouper family that have abandoned their bottom dwelling lifestyle in favour of mid-water plankton feeding. Their dense schools are an integral part of the Indo-Pacific reef community. They are capable of sex reversal from female to male.

Above top: Yellowline Basslet, *Tosanoides flavofasciatus* (10 cm), in 52 m, Oshima I., Japan.
Above bottom: Cherry Blossom Basslet, *Sacura margaritacea* (10 cm), in 22 m, Osezaki, Japan.
Opposite top: Redstripe Basslet, *Pseudanthias fasciatus* (12 cm), in 20 m, Loloata I., Papua New Guinea.
Opposite bottom: Luzon Basslet, *Pseudanthias luzonensis* (10 cm), in 20 m, Rinca I., Indonesia.

Above and opposite: Mating Crinoid Cuttlefish face off, then suddenly come together and lock arms. While their arms are intertwined, the male transfers sperm packets to the female. After mating they appear to mimic a pair of sharp-nosed puffers, a toxic fish ignored by most predators.

Crinoid Cuttlefish, *Sepia* sp. (5 cm), in 20 m, Lembeh Strait, Sulawesi, Indonesia.
(photo opposite top by S. Michael)

Above and opposite: This diverse collection of sea slugs features some of the lesser known animals broadly classed by divers as nudibranchs. They represent both herbivore and carnivore that may or may not have a bubble or internal shell to protect their visceral organs.

Above top: Souverby's Lobiger, *Lobiger souverbii* (4 cm).
Above bottom: Psychedelic Slug, *Sagaminopteron psychedelicum* (1 cm).
Opposite top: Green Angel, *Notobryon* sp. (3 cm).
Opposite bottom: Green Oxynoe, *Oxynoe viridis* (3 cm), in 4-12 m, Milne Bay, Papua New Guinea.

The cryptic Comet lives within cracks and caves in the reef. As seen here, they will lay their eggs on the ceiling of a crevice. The male will fan them with its fins and chase off egg predators until hatching.

Comet, *Calloplesiops altivelis* (18 cm), in 10 m, Lembeh Strait, Sulawesi, Indonesia.

One of the most dramatic cases of mimicry among marine animals involves certain pygmy angelfishes and juvenile surgeonfishes. The mimicry of this particular species pair is so complete that the surgeon even has an amazing blue painted replica of the angel's protective cheek spine.

Top: Lemonpeel Angelfish, *Centropyge flavissima* (9 cm), in 15 m, Christmas I., Indian Ocean.
Bottom: Mimic Surgeonfish, *Acanthurus pyroferus* (8 cm), in 12 m, Port Vila, Vanuatu.

Sand anemone

Ray

Jawfish

Snake eel

Crinoid

Mantis shrimp

Cuttlefish

Lionfish

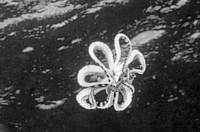
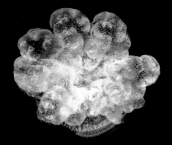

Seasnake

Jellyfish

Starfish

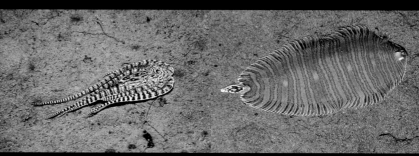

Flounder

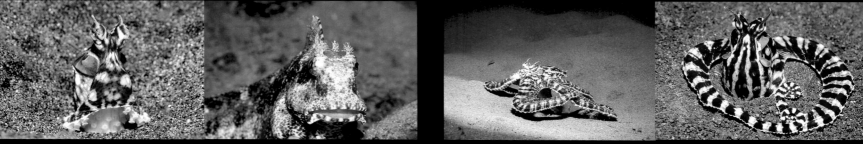

Blenny

Undetermined

Opposite and above: The incredible Mimic Octopus can readily change shape (and to a limited degree, hues and markings) to appear as a variety of animals. It copies the exact movements of the subjects mimicked to enhance its deception. Its repertoire also includes the guise of a nudibranch, hermit crab, seahorse, adhesive anemone and ghost crab.

Mimic Octopus, *Thaumoctopus mimicus* (30-60 cm), in 4-15 m, Flores, Sulawesi and Bali, Indonesia. (Mimic jellyfish photo D. Neilsen-Tackett)

This squid has just captured a cardinal fish and swum to the surface. A gelatinous grey egg mass from the mouth of the cardinal can be seen oozing from within the tentacles as a result of the holding pressure applied by the squid.

Bigfin Reef Squid, *Sepioteuthis lessoniana* (30 cm), Milne Bay, Papua New Guinea.

The White-saddled Shrimp is a tiny crustacean that exhibits unique behaviour. It lifts the tail and stands almost vertically, waving its body. Mostly found in groups clustered on or near certain corals and anemones, it occurs in all tropical waters.

White-saddled Shrimp, *Thor amboinensis* (1-1.5 cm), in 12 m, Lembeh Strait, Sulawesi, Indonesia.

Above and opposite: Nowhere on earth is there such a stark contrast between terrestrial and marine environments than along the Red Sea coast. Barren deserts abruptly give way to an explosion of marine life amongst a pristine seascape.

Sharm-el-Sheikh, Egypt.

Above and opposite: Snake eels are perfectly adapted to open sandy areas. They live under the substrate and are often seen with only the head, eyes, or snout exposed. They are efficient scavengers and voracious carnivores, ever watchful for passing prey.

Spoon-nose Eel, *Ichiophis intertinctus* (1 m), in 10 m, St. Vincent, Caribbean.

Opposite and above: Sand divers are ideally adapted to life in the undersea desert. They dart into the substrate when threatened and can literally swim through the sand. In the same way the bulbous bow of a tanker more efficiently goes through water, the enlarged lower jaw of certain sand divers may help them push through the sand.

Opposite top: Goldbar Sand Diver, *Trichonotus halstead* (15 cm), in 6m, Lembeh Strait, Sulawesi, Indonesia.
Opposite bottom: Spotted Sand Diver, *Trichonotus setiger* (12 cm).
Above: Unidentified Sand Diver, *Trichonotus* sp. (22 cm), in 10-12 m, Milne Bay, Papua New Guinea.

Above and opposite: The sponges are filter feeders that are constantly pumping huge volumes of water to strain out microscopic prey. Small holothurians gather on the surface to take advantage of the detritus left behind during the feeding process.

Barrel Sponge, *Xestospongia testudinaria* (80 cm); Social Holothurian, *Synaptula lamperti*, in 15 m, Russell Is, Solomon Is.

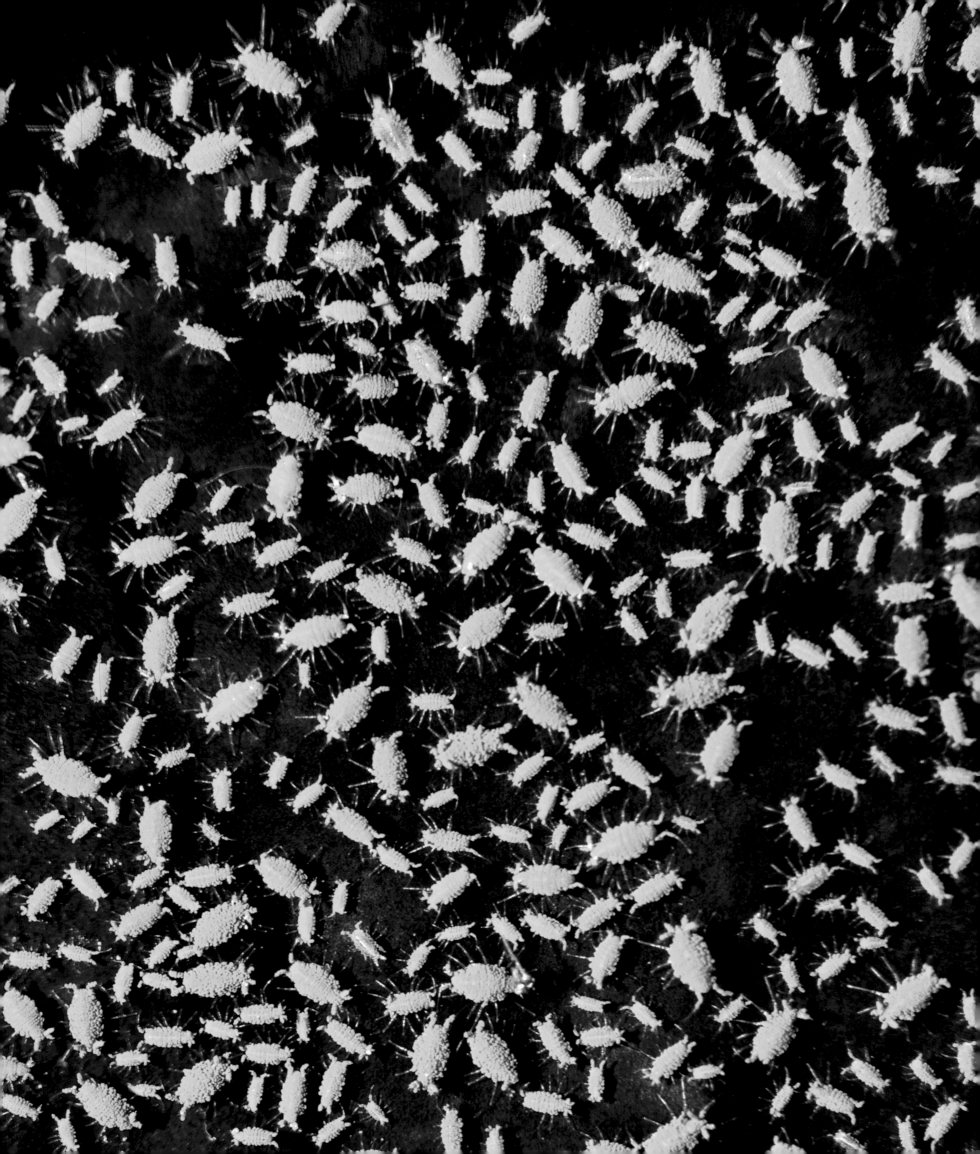

Above: The Harlequin Sweetlips undergoes an amazing colour change as it grows. It also exhibits a change in behaviour, from a hyperactive, undulating juvenile, to a more relaxed adult. A "teenage" individual is seen here.

Harlequin Sweetlips, *Plectorhinchus chaetodontoides* (20 cm), in 12 m, Lembeh Strait, Sulawesi, Indonesia.

Opposite: There are numerous marine invertebrates about which little is known. These tiny yellow isopods occur on sponges in dense aggregations sometimes numbering in the thousands. This family is widespread in both temperate and tropical seas.

Swarming isopods, *Santia* sp. (2 mm), in 12 m, Milne Bay, Papua New Guinea.

Above and opposite: In most locations, all but the inner pectoral fins of the Spiny Devilfish are drab. However in Indonesia's Lembeh Strait these fish exhibit an amazing array of colours overall. Scientists are still not sure why this is the case.

Spiny Devilfish, *Inimicus didactylus* (12-22 cm), in 5-20 m, Lembeh Strait, Sulawesi, Indonesia.

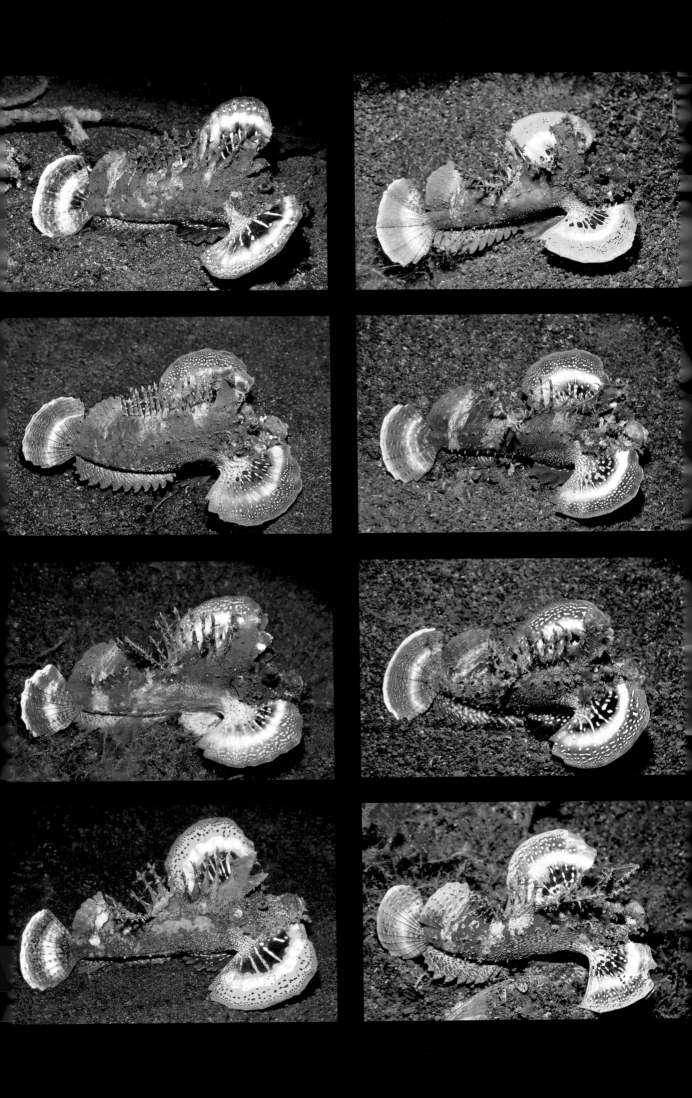

Above and opposite: The incredible symbiotic relationship between a stomatopod and decapod is documented in these photographs. The cleaner shrimp guests receive food scraps and protective shelter from their host, while the mantis has a permanent cleaning service. Several generations of shrimp can occupy the burrow during the lifespan of the mantis.

Cleaner shrimp, *Stenopus tenuirostris* (4 cm); Giant Mantis Shrimp, *Lysiosquillina lisa*, in 10 m, Milne Bay, Papua New Guinea.

Above : The juvenile Spotted Drum hides under ledges and in caves during the day. It is a crustacean predator that does most of its feeding after dark. The spectacular flowing fins reduce as the fish matures.

Spotted Drum, *Equetus punctatus* (10 cm), in 10 m, Bonaire, Caribbean.

Opposite: This apparently undescribed aeolid nudibranch is associated with gorgonian sea fans. Aeolids lack distinct gills and utilize cerata, the tentacular projections on the body, for respiration and defense.

Seafan Slug, *Phyllodesmium* sp. (4 cm), in 20 m, Lembeh Strait, Sulawesi, Indonesia.

Opposite: Sundials have distinctive markings and texture that allows easy recognition. They are found as deep as 300 m. and feed exclusively on burrowing coelenterates including sea pens and anemones. The photos capture their behaviour as a prelude to mating.

Sundial, *Architectonica perspectiva* (4 cm), in 12 m, Milne Bay, Papua New Guinea.

Above: This species of Headshield Slug stalks its prey by tracking the mucous trail of other opisthobranchs utilizing eyes and sensory appendages near the mouth. They live on sandy substrates and group gatherings like this probably relate to reproductive activity.

Headshield Slug, *Chelidonura hirundinina* (3 cm), in 10 m, Secret Bay, Bali, Indonesia.

Above: The algal moustache worn by this porcupinefish has probably resulted from a wound. Fish skin is normally covered with slime that has antibiotic properties. If the tough connective tissue under the skin is exposed, it makes a suitable base for algal spores to establish.

Fine-spotted Porcupinefish, *Diodon holocanthus* (20 cm), in 5 m, Lembeh Strait, Sulawesi, Indonesia.

Opposite: Chance encounters such as this provide a wonderful photographic opportunity. This snake eel sports the latest in bizarre headgear. A *Cassiopea* jellyfish has inadvertently settled onto the eel burrow, creating this outrageous fashion statement.

Upside-down Jellyfish, *Cassiopea* sp. (8 cm); Snake eel, ophichthid sp., in 10 m, Secret Bay, Bali, Indonesia.

Above: A number of fishes have structures used to entice prey into striking range. As well as the better known anglerfishes, the Stargazer has long filaments inside the lower jaw that look like a writhing mass of polychaete worms when it spits them above its head.

Stargazer, *Uranoscopus sulphureus* (30 cm), in 12 m, Lembeh Strait, Sulawesi, Indonesia.

Opposite: Not only do the anglerfishes rely on luring to capture their prey, in many cases they simply ambush or stalk their quarry. The long skin filaments on this angler species helps it disappear among algae covered substrate.

Striated Frogfish, *Antennarius striatus* (15 cm), in 5 m, Lembeh Strait, Sulawesi, Indonesia.

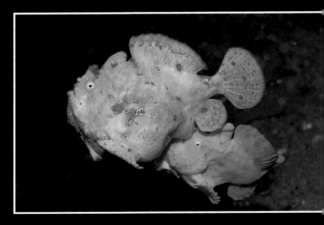
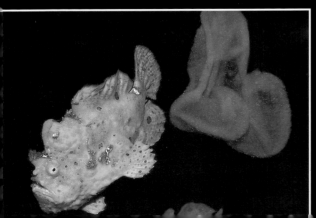
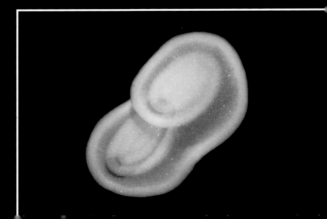

Opposite: Reproduction in anglerfish is a tender affair. On spawning day, the male will occasionally caress the female with his hand-like pectoral fins. After dark they rise in the water column and it appears the male induces the egg raft from the female's cloaca with his mouth.

Painted Frogfish, *Antennarius pictus* (12 cm), in 6 m, Lembeh Strait, Sulawesi, Indonesia.

Above: The hamlets have both functional male and female sex organs. During the evening spawning period, individuals will trade roles, acting as the male during one bout and as the female in the next. Hamlet spawning is easily observed as the sun begins to set.

Butter Hamlet, *Hypoplectrus unicolor* (7 cm), in 12 m, Bonaire, Caribbean.

Some of the head markings on reef fishes, like this grubfish and angelfish, are reminiscent of the facial tattoos worn by New Zealand Maoris. In fishes, these markings help individuals recognize members of their own kind.

Above: Neon Grubfish, *Parapercis pulchella* (12 cm), in 18 m, Osezaki, Japan.
Opposite: Blue-stripe Angelfish, *Chaetodontoplus septentrionalis* (15 cm), in 10 m, Osezaki, Japan.

Some bottom-dwelling fishes have developed wing-like pectoral fins. These exaggerated structures may serve several functions, one being to increase the apparent size of the fish so that it looks too large for a predator to eat.

Above top: Flying Gurnard, *Dactylopterus volitans* (20 cm), in 15 m, St. Vincent, Caribbean.
Above bottom: Longtail Waspfish, *Apistus carinatus* (12 cm), in 12 m, Osezaki, Japan.
Opposite top: Bluefin Lionfish, *Parapterois heterura* (10 cm), in 10 m, Secret Bay, Bali, Indonesia.
Opposite bottom: Sea Moth, *Eurypegasus draconis* (7 cm), in 10 m, Lembeh Strait, Sulawesi, Indonesia.

Above: Ghost shrimps are superabundant in some muddy habitats, but are seldom seen by divers. They rarely leave the security of their burrows, but occasionally are seen at night in the open. They are a favourite bait for anglers.

Ghost Shrimp, *Callianassa* sp. (5 cm), in 15 m, Lembeh Strait, Sulawesi, Indonesia.

Opposite: This decorator crab has attached red algae to bristles on its body to break up its outline and make it less prone to predation. Several other crab species utilize natural decorations in a similar manner.

Spider Crab, *Archaeus* sp. (4 cm), in 18 m, Milne Bay, Papua New Guinea.

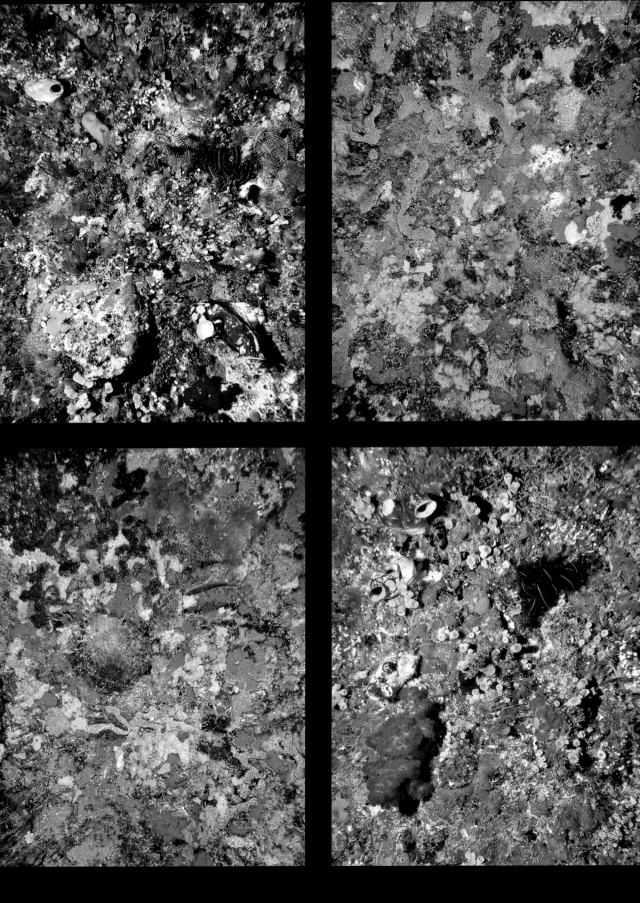

Above and opposite: Nature's palette paints an endless combination of colours. Available space is quickly colonized by an array of competing organisms forming a patchwork quilt of contrasting shades.

Mixed invertebrates, unidentified, in 5-8 m, St.Vincent, Caribbean; Lembeh Strait and Raja Ampat s., Indonesia.

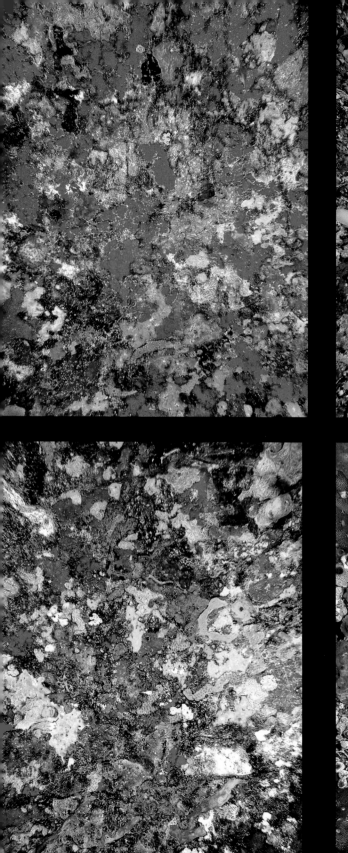

Above: The sharpnose puffers or tobies look quite defenseless, but possess several forms of defence. They can inflate when attacked and their skin and organs are permeated with deadly toxins. Predators soon learn to avoid them

Crowned Toby, *Canthigaster coronata* (5 cm), in 18 m, Tulamben, Bali, Indonesia.

Opposite: Corallimorpharians cluster together to form a comical facial expression not unlike Homer Simpson. These organisms, which are related to anemones, appear as flattened discs, but can form balloon-like structures as shown here.

Balloon Corallimorph, *Amplexidiscus fenastrafer* (30 cm), in 8 m, Milne Bay, Papua New Guinea.

Many of the larger angelfishes undergo dramatic colour changes as they grow. This group of photos documents the chromatic metamorphosis that occurs in two similar Atlantic angels. These species are so closely related that they regularly interbreed.

Above top: Blue Angelfish, *Holocanthus bermudensis* (4 cm),
Above bottom, and opposite: Queen Angelfish, *Holocanthus ciliaris* (10-35 cm),
in 10-15 m, Florida Keys, USA; Bonaire and St. Vincent, Caribbean.

Fragile corals thrive best in waters that are protected from the influence of wind and waves.
These conditions usually prevail on the lee side of islands or in lagoons. The stable environment

Hydroids are cousins of corals, sea anemones, and jellyfish and like their cnidarian relatives possess specialized stinging cells. The Magnificent Hydroid has a single cluster of tentacles at the end of a stalk, unlike most others that have feathery branches.

Magnificent Hydroid, *Ralpharia magnifica* (10 cm), in 2 m, Port Philip Bay, Victoria, Australia.

The Elegant Shrimp is a smaller relative of the renowned Harlequin Shrimp. It occurs in tropical waters of the Indo-Pacific, but is relatively unknown due to its cryptic habits. It is an echinoderm predator like its better known relative.

Elegant Shrimp, *Phyllognathia ceratophthalma* (1-1.5 cm), in 20 m, Lembeh Strait, Sulawesi, Indonesia.

Above and opposite: In vivid contrast to the gargantuan deepwater squids, the Class Cephalopoda also includes a host of smaller members. Some, such as the three illustrated squids, rarely exceed 10 cm. Others grow considerably larger, including the adult of the tiny orange and white cuttlefish featured here that will grow to 50 cm.

Above top: Broadclub Cuttlefish, *Sepia latimanus* (3 cm), in 12 m, Lembeh Strait, Sulawesi, Indonesia.
Above bottom: Striped Pyjama Squid, *Sepioloidea lineolata* (4 cm), in 5 m, Edithburgh, South Australia.
Opposite top: Bobtail Squid, *Euprymna berryi* (3 cm), in 10 m, Lembeh Strait, Sulawesi, Indonesia.
Opposite bottom: Tropical Bottletail Squid, *Sepiadarium kochi* (2 cm), in 8 m, Milne Bay, Papua New Guinea.

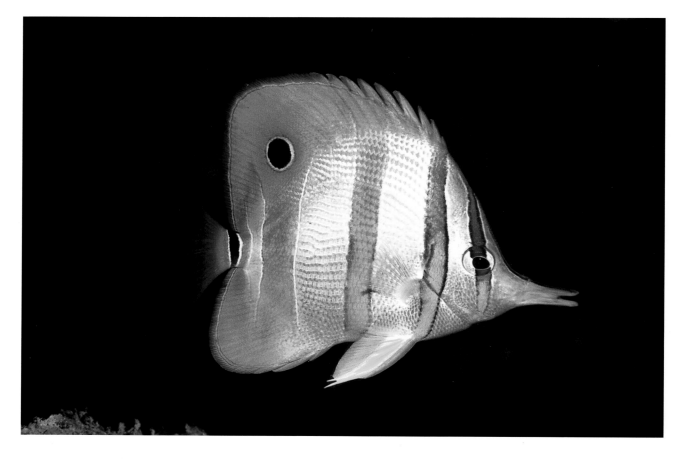

A number of butterflyfishes have long snouts that enable them to feed on organisms between coral branches and in reef crevices. The Beaked Coralfish preys heavily on tube worm appendages, while the Saddled Butterflyfish feeds mainly on filamentous algae, hard-coral polyps, sponges and tunicates.

Top: Beaked Coralfish, *Chelmon rostratus* (10 cm), in 12 m, Bintan Is, Indonesia.
Bottom: Saddled Butterflyfish, *Chaetodon ephippium* (15 cm), in 12 m, Milne Bay, Papua New Guinea.

The white flower-like organisms attached to this buried sea pen are known as stauromedusans. Despite their appearance, they are actually tiny, non-swimming jellyfish. They are found on open sand, but rarely seen.

Attached jellyfish, possibly *Lipkea* sp. (1 cm), in 18 m, Tulamben, Bali, Indonesia.

The highly ornate and conspicuous feather duster and Christmas tree worms are common on most tropical reefs. Young worms settle on coral heads and secrete a tube that kills the underlying polyps. They are found individually, in small groups, or in tight clusters.

Opposite: Social Feather Duster, *Bispira brunnea* (2 cm), in 12 m, St. Vincent, Caribbean.
Above: Serpulid worm, possibly *Spirobranchus* sp. (8 cm), in 10 m, Komodo, Indonesia.

Above, opposite and overleaf: Seaweed (algae) is universally common, but almost always under
appreciated. It plays a vital role in all marine ecosystems. A huge range of size and structure is
represented, from miniscule microscopic species to giant kelp, which may grow to 50 metres
length. In 3-12 m, Recherche Is., Western Australia; Spencer Gulf and St. Vincent Gulf, South
Australia; Milne Bay, Papua New Guinea; and Raja Ampat Is,. Indonesia.

Opposite: Mermaid's Wine Glass, *Acetabularia major* (1 cm), in 6 m, Anilao, Philippines.
Above: Pom Pom Alga, *Tydemania expeditionis* (50 cm), in 12 m, Milne Bay, Papua New Guinea.

Vesicle Caulerpa, *Caulerpa vesiculifera* (7 cm)

Coin Caulerpa, *Caulerpa nummularia* (5 cm)

Sargassum Weed, *Sargassum* sp. (12 cm)

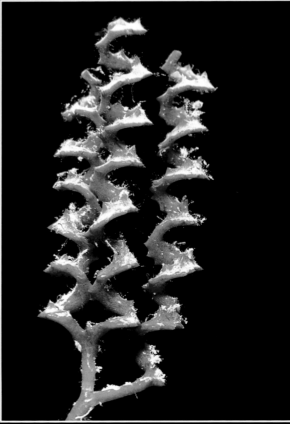

Corkscrew Alga, *Caulerpa serrulata* (6 cm)

Turbinweed, *Turbinaria ornata* (10 cm)

Cactus Caulerpa, *Caulerpa cactoides* (8 cm)

Pom Pom Alga, *Tydemania expeditionis* (8 cm)

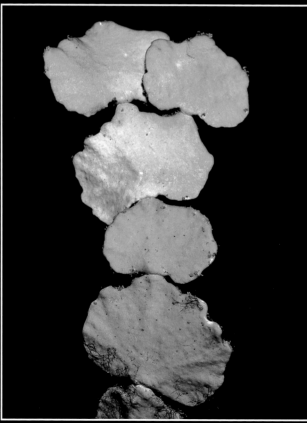

Coralline Alga, *Halimeda macrolaba* (8 cm)

Sea urchins and holothurians (sea cucumbers) are two of the major groups of echinoderms commonly seen in shallow waters. They use their modified oral tube feet to feed on organic bottom deposits. Some urchins and sea cucumbers are used for human consumption, particularly in the Orient.

Above : Bell's Sea Urchin, *Salmacis belli* (8 cm); Yellow holothurian, *Cucumaria miniata*, in 15 m, Komodo, Indonesia.
Opposite: Graeff's Sea Cucumber, *Bohadschia graeffei* (10 cm), in 5 m, Milne Bay, Papua New Guinea.

Blennies, especially the tubeblennies, spend much of their time with only their heads projecting from a hiding place. Thus, it should not be surprising that the males of some species have head ornamentation. The "head dress" may serve to advertise their presence to consexual rivals or potential mates.

Above: Tubeblenny, *Neoclinus bryope* (5 cm), in 12 m, Osezaki, Japan.
Opposite: Tasmanian Blenny, *Parablennius tasmanianus* (7 cm), in 5 m, Edithburgh, South Australia.

Shrimp gobies are well known for the symbiotic association they have with pistol shrimps. But in some areas, like Osezaki, Japan, the goby and shrimp regularly share their burrow home with the Longtail Dartfish.

Japanese Shrimp Goby, *Amblyeleotris japonica* (8 cm); Longtail Dartfish, *Ptereleotris hanae*, in 15 m, Osezaki, Japan.

Flatfishes start their lives looking like any other fish, however as they mature one eye migrates over the top of the head until both eyes are on the same side. This makes sense for a fish that spends its life lying on its side.

Top: Unidentified flatfish, (3 cm), in 1 m, Great Barrier Reef, Australia.
Bottom: Flounder, *Bothus* sp. (20 cm), in 5 m, Lembeh Strait, Sulawesi, Indonesia.

The beautiful Harlequin Shrimp can exhibit somewhat grizzly feeding behaviour. They have been seen to flip seastars over and feed on the end of the arms, working their way towards the central disc. In this way, they can make their living buffet last longer, and also prevent its escape.

Above and opposite: Harlequin Shrimp, *Hymenocera picta* (4 cm), in 18 m, Lembeh Strait, Sulawesi, Indonesia.

The Hingebeak Shrimps are a gregarious species that can be found in extremely large numbers. Although these shrimps are not known to engage in cleaning behaviour, aggregations often reside at a cleaning station.

Durban Hingebeak Shrimp, *Rhynchocinetes durbanensis* (4 cm), in 20 m, Tulamben, Bali, Indonesia.

Opposite: The female rainbow mantis shrimp provides a mobile nursery. Her eggs are not attached and she periodically repositions and adjusts the egg blanket, probably a strategy that ensures adequate oxygenation and cleansing. Other mantis species place their eggs in a burrow.

Rainbow Mantis Shrimp, *Odontodactylus scyllarus* (15 cm), in 8 m, Milne Bay, Papua New Guinea.

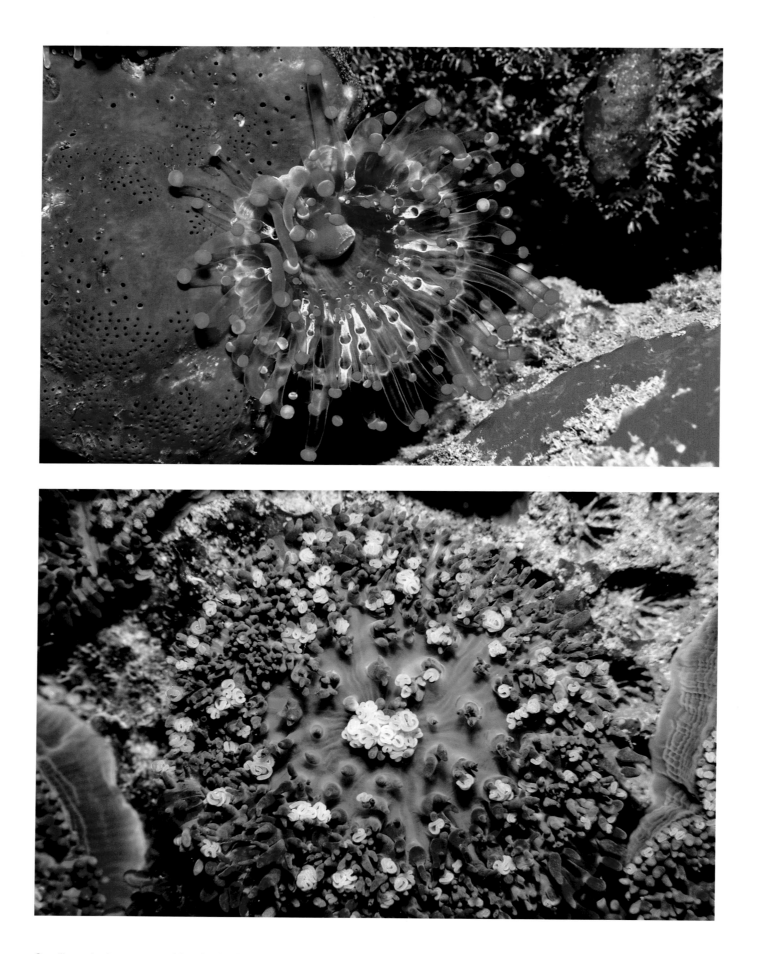

Corallimorpharians are considered to be intermediate between hard corals and sea anemones. Like other cnidarians, they have stinging nematocyst cells, used to capture planktonic food. Some are also armed with stinging threads (acontia), which may be dangerous to divers.

Top: Orange Ball Corallimorph, *Pseudocorynactis caribbeorum* (8cm), in 8 m, St. Vincent, Caribbean.
Bottom: Biscuit Corallimorph, *Discosoma* sp. (5 cm), in 4 m, Lembeh Strait, Sulawesi, Indonesia.

This striking frogfish is probably a species unknown to science. It was discovered by anglerfish authority Scott Michael in Lembeh Strait, Indonesia. Worldwide there are currently over 40 valid species recognized. This one bears similarities to the Striated Frogfish but has a different esca (lure).

Lembeh Frogfish, *Antennarius* sp. (10 cm), in 5 m, Lembeh Strait, Sulawesi, Indonesia.

Parasitism is a type of symbiosis (living together) involving a wide range of organisms. One species, the parasite, lives in or on its host and obtains nourishment, causing harm. Other types of symbiosis involve harmless relationships, such as the barnacle-crayfish association seen here.

Above, top and bottom: Parasitic isopods, unidentified (1-2 cm), in 10 m, Loloata I., Papua New Guinea and Bonaire, Caribbean.
Opposite top: Starfish snail, *Thyca* sp. (5 mm), in 20 m, Bintan I., Indonesia.
Opposite bottom: Barnacles, unidentified (3 mm), in 10 m, Milne Bay, Papua New Guinea.

305

The polyclad flatworm (above) has a toxic mucous coating. The young Pinnate Batfish (opposite) adopts these colours and can also swim on its side to look like the poisonous flatworm, a strategy to deter predatory attack. At least one species of flatfish is also known to duplicate this mimicry.

Above: Flatworm, *Pseudobiceros hymanae* (7 cm), in 10 m, Russell Is, Solomon Is.
Opposite: Pinnate Batfish, *Platax pinnatus* (4 cm), in 6 m, Milne Bay, Papua New Guinea.

Dendrodoris tuberculosa (10cm)

Phyllidia ocellata (6 cm)

Mexichromis multituberculata (5 cm)

Glossodoris stellata (5 cm)

Hypselodoris sp. (5 cm)

Ceratosoma magnifica (7 cm)

Tambja sp. (7 cm)

Chromodoris kuniei (5 cm)

Sea slugs, the butterflies of the sea, are a primary objective of underwater photographers. They range widely across temperate and tropical seas, displaying considerable variability in size, shape, and colour. These striking molluscs are often found in close proximity to their food source (e.g. sponges, algae, and hydroids). Photographed in 3-20 m, Secret Bay and Lembeh Strait, Indonesia; Madang and Milne Bay, Papua New Guinea.

Chromodoris magnifica (4 cm)

Flabellina bilas (6 cm)

Gymnodoris rubropapulosa (6 cm)

Ardeadoris egretta (6 cm)

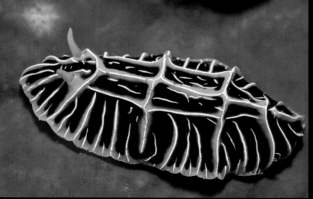

Reticulidia halgerda (5 cm)

Petalifera ramosa (4 cm)

Cerberilla ambonensis (5 cm)

Flabellina rubrolineata (4 cm)

On rare days the normally turbulent ocean transforms to mill-pond conditions. Placid surface waters mirror the cavalcade of life below only when this occurs. If this situation persists for a lengthy period, the lack of circulation and consequent oxygen depletion may prove fatal to some reef creatures.

Surface reflections, above, Palau; and opposite, Milne Bay, Papua New Guinea.

Above and opposite: There are a number of toadfishes in the Caribbean that have very restricted ranges. This is because they do not have a pelagic larval stage, but instead the parents care for the newly hatched young. The Splendid Toadfish is only found off the island of Cozumel.

Splendid Toadfish, *Sanopus splendidus* (20 cm), in 15 m, Cozumel, Mexico.

Divers rarely see the young of the Sailfin Snapper. Attaining an adult size of 60 centimetres, this species is restricted to the West Pacific and has been observed in schools of hundreds of individuals in deep areas of sand and rubble.

Sailfin Snapper, *Symphorichthys spilurus* (10 cm), in 3 m, Lembeh Strait, Sulawesi, Indonesia.

Many of the smaller fishes that live on the sea floor defend limited resources, like food or shelter. In this case, a tubeblenny displays vigorously in an effort to appear menacing to an intruder.

Tubeblenny, *Neoclinus bryope* (6 cm), in 2 m, Osezaki, Japan.

Nocturnally active *Saron* shrimps are among the most attractive animals to be seen underwater. Although scientists disagree on the exact number, there are relatively few species. Some with uncertain status may prove to be only geographical variants.

Marbled Shrimp, *Saron* sp. (10 cm), in 5 m, Milne Bay, Papua New Guinea.

Pebble crabs are common residents of Indo-Pacific reefs, occurring intertidally to a depth of at least 60 m. They are usually found on sand or sand-mud substrates, but several species inhabit coral reefs. The carapace typically forms a protruding snout-like structure.

Painted Pebble Crab, *Leucosia anatum* (3 cm), in 8 m, Milne Bay, Papua New Guinea.

Ascidians occur as solitary individuals or as colonies, presenting a bewildering range of shapes and colours. Colonies may include thousands of individuals, and exhibit considerable intraspecific variation. They are one of the predominant filter-feeding groups.

Opposite: *Polycarpa aurata* (8 cm), in 5 m, Raja Ampat Is., Indonesia.
Above: *Didemnum* spp., *Clavelina* sp., *Rhopalaea* sp. (colony 12 cm), in 12 m, Tulamben, Bali, Indonesia.

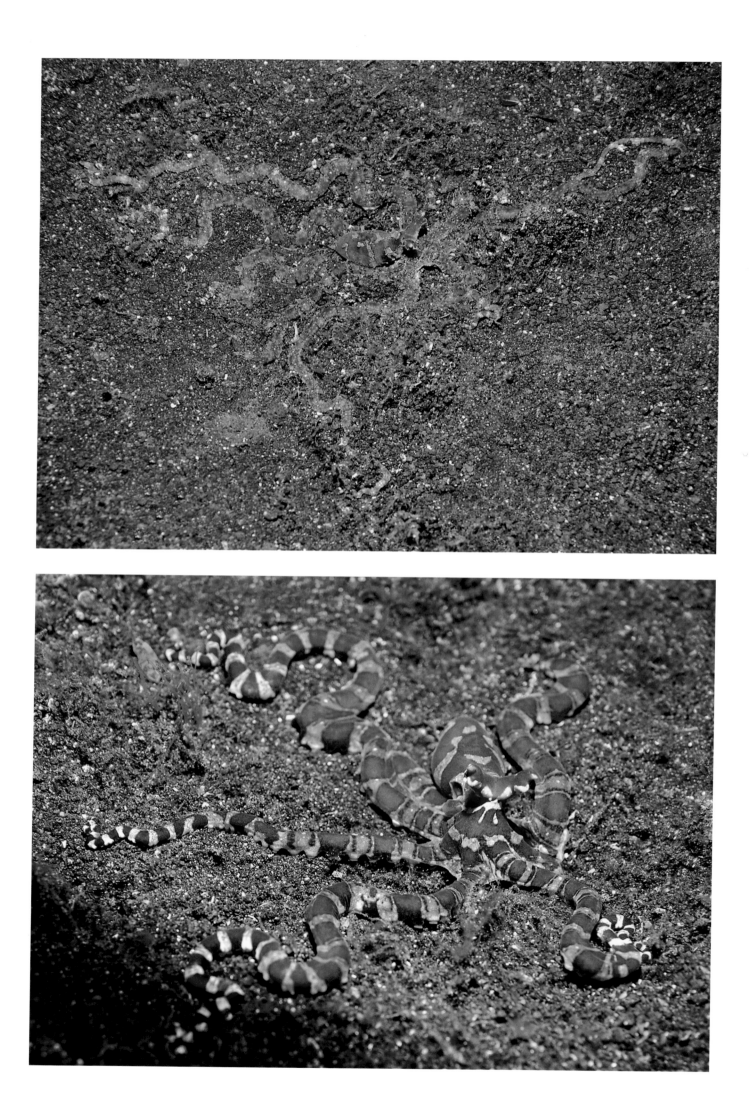

Above and opposite: The amazing ability of an octopus to dramatically alter its appearance is vividly illustrated in this photographic sequence taken within a 20-second time frame. This colourful species has only come to prominence in recent times and still lacks an official scientific name.

Ornate Octopus, *Octopus* sp. (45cm), in 6 m, Lembeh Strait, Sulawesi, Indonesia.

Above and opposite: Golden Turret Corals are generally found in caves or under overhangs. This colony,
protruding from a silty bottom, was part of an extensive aggregation featuring unusually elongated
skeletal structures and polyps that reached an unusual 8 cm in diameter.

Turret Coral, *Tubastrea* sp. (20 cm), in 30 m, Anilao, Philippines.

The wide-ranging Linecheek Wrasse is a beautiful inhabitant of coral reefs in the Indo-West Pacific from the Red Sea to Micronesia. It has been observed changing colour to inconspicuously join a school of goatfish from where it successfully launched attacks on unsuspecting smaller fish.

Linecheek Wrasse, *Oxycheilinus diagrammus* (20 cm), in 10 m, Anilao, Philippines.

Most often seen in part only protruding from a hole in the substrate, this eel will occasionally venture away from the safety of its burrow in search of prey. Its unique swimming action is reminiscent of a rhythmic gymnastic ribbon display.

Ribbon Eel, *Rhinomuraena quaesita* (100 cm), in 18m, Anilao, Philippines.

Frogfishes have modified pectoral and pelvic fins that are used much like hands and feet. They come in all colour forms from plain to gaudy and are generally camouflaged to blend in with their surroundings, particularly reef, rubble, plants and sponges.

Above and opposite: Warty Frogfish, *Antennarius maculatus* (10 cm), Painted Frogfish, *Antennarius pictus* (30 cm), in 3-15 m, Anilao, Philippines.

The exotic Flasher Wrasses are most conspicuous when courting males erect their fins and turn on intense luminous colours to perform a frantic ritualistic dance to attract gravid females. When displaying, they become the most difficult underwater photographic target possible.

Opposite top: Carpenter's Flasher, *Paracheilinus carpenteri* (7 cm).
Opposite bottom: Linespot Flasher, *Paracheilinus lineopunctatus* (7 cm).
Above: Hybrid Flasher, *Paracheilinus* sp. (7cm), in 20-25 m, Anilao, Philippines.

Above: Most allied cowries are small and spend their entire lives on one large fan or octocoral. This large species, reaching 12 cm, is one exception that roams the sand or mud substrate to seek and feed on smaller soft corals.

Shuttlecock Egg Cowry, *Volva volva* (8 cm), in 22m, Anilao, Philippines.

Opposite: The attractive pattern, shapes and tonings of this hamburger-size sea urchin belies the fact it has extremely venomous pedicellaria that can inflict serious stings. Two fatalities have been reported from Japan by divers holding them against their bodies while collecting urchins for consumption.

Toxic Urchin, *Toxopneustes pileolus* (10 cm), in 10 m, Anilao, Philippines.

Perhaps the most successful hunters on coral reefs, the Lizardfishes use camouflage and ambush tactics to launch upward attacks at prey. They are fierce predators that have been known to strike at fishing lures as big as themselves.

Lizardfish, *Synodus* sp. (22 cm), in 8 m, Anilao, Philippines.

The furry growth evident on this scorpionfish is actually a covering of live hydroids. This recently discovered phenomenon is not fully understood. Perhaps the fish encourages the growth to enhance its camouflage capabilities. Some Frogfishes are also known to harbour these coelenterates.

Painted Stingfish, *Minous pictus* (7 cm), in 18 m, Anilao, Philippines.

Opposite: Like their relatives the anemones, corals have tentacles but possess milder stinging nematocysts. Most have small polyps, but those with larger tentacles are effectively like an anemone and offer a home to certain organisms that have immunity to the sting.

Opposite top: Mushroom Coral Pipefish, *Siokunichthys nigrolineatus* (10 cm)
Opposite bottom: Commensal Shrimp, *Periclimenes kororensis* (3 cm), in 10-12 m, Anilao, Philippines.

Above: The Dartfishes are a group of small, elegant gobies that have upturned mouths. They hover above the bottom alone, in pairs or small aggregations to catch passing zooplankton and quickly retreat into burrows when approached.

Decorated Dartfish, *Nemateleotris decora* (6 cm), in 25 m, Milne Bay, Papua New Guinea.

Above: The gill structure of most tropical nudibranchs is often reduced and the organ seldom stands totally above the body. One exception is this recently named highly coloured species whose massive gills can be as thick as a pencil and 8 cm across.

Lavender Slug, *Hypselodoris apolegma* (12 cm), in 10 m, Anilao, Philippines.

Opposite: These tunicates are filter feeders that obtain nourishment by pumping seawater through a U-shaped internal system that has both inlet and outlet. They have the ability to close these incurrent and excurrent openings in response to external disturbances.

Tunicates, *Rhopalaea* sp. (2 cm), in 18 m, Anilao, Philippines.

Photographic Details

The photos in this book were taken with 100 ASA film utilizing 14 mm, 20 mm, 28-80 mm, 105 mm, and 70-185 mm lenses. All were shot with a single top-mount flash except the occasional wide angle. The 70-185mm has a 2X wet diopter attachment for close-ups and is the lens I use most. The microscopic pics. were taken on an optical bench with tungsten film.

I invite readers to make contact with information on behaviour, observations, discoveries or nomenclature updates of marine life.

Roger Steene
Box 188
Cairns 4870
Australia